DICTIONARY OF NANOTECHNOLOGY

DICTIONARY OF NANOTECHNOLOGY

Edited by

Thomas George

ANMOL PUBLICATIONS PVT. LTD.
NEW DELHI - 110 002 (INDIA)

ANMOL PUBLICATIONS PVT. LTD.
H.O.: 4374/4B, Ansari Road, Darya Ganj,
New Delhi-110 002 (India)
Ph.: 23261597, 23278000

B.O.: No. 1015, Ist Main Road, BSK IIIrd Stage
IIIrd Phase, IIIrd Block,
Bangalore - 560 085 (India)
Visit us at: www.anmolpublications.com

Dictionary of Nanotechnology

First Published, 2006

PRINTED IN INDIA

Printed at Mehra Offset Press, Delhi.

Preface

Nanoscience and nanotechnology are recent advancement in the field of science and technology. According to one projection, by 2010 technology will be 2000 times more advanced than in 1900. In fact, 'nano' is used to describe a billionth of a unit, so nanometer is a billionth of a meter.

But nanoscience is not just microtechnology on a smaller scale; it is much more than a mere improvement on previous forms of technology. Nanoscience is a completely new production process in a molecular world, opening up possibilities far beyond what we have known before. Nature provides the inspiration for this new technology—from time immemorial, nanomachines have been at work in all organic cells, building our world of plants, animals and people from a store of chemical elements.

Researches in nanotechnology have great impact on existing knowledge. And a revolution is expected in all fields of science—medicine, physics, chemistry, biotechnology etc. in near future.

If used in constructive way, the miraculous nanotechnology will prove a boon in the service of humanity and life on the earth. Especially in the field of medicine and surgery, nanotechnology can bring tremendous changes. Many dreaded diseases can be cured which are now only preventable.

This dictionary compiles major relevant terms pertaining to the nanoscience/nanotechnology. The description and definitions are borrowed from different authoritative sources to make it a dependable reference tool. The editor owes his gratefulness to all those whose works are cited or substantially made use of.

Suggestions and constructive criticisms from all quarters are welcome which will help improving the quality and utility of the work.

—Editor

A

ABC transporters: Members of a large superfamily of membrane transport proteins that hydrolyze ATP and transfer a diverse array of small molecules across a membrane.

Abrasive: A hard and wear-resistant material (such as a ceramic) used to wear, grind, or cut away other material.

Absolute Zero: The physical limit to how low a temperature any matter can have, which equals –273.15° C.

Abstraction reaction: A reaction that removes an atom from a structure.

Accuracy: The degree of correctness with which the measuring system yields the "true value" of a measured quantity, where the"true value" refers to an accepted standard, such as a standard meter or volt. Typically described in terms of a maximum percentage of deviation expected based on a full-scale reading.

ACE Paste: Atomspheric Carbon Extractor. Harvests the greenhouse gases for Carbon, to be used for diamondoid fabrication. Larger than most pastebots, because it has to be collectible afterwards. A well-designed paste could harvest 100X or more its empty weight. ACE Paste may not be necessary, because large fixed installations might be more efficient.

Acetyl CoA: Small water-soluble molecule that carries acetyl groups in cells. Comprises an acetyl group linked to coenzyme A (CoA) by an easily hydrolyzable thioester bond.

Acetyl: Chemical group derived from acetic acid. Acetyl groups are important in metabolism and often added as a covalent modification of proteins.

Acetylcholine receptor: Ion channel that opens in response to acetylcholine binding, thereby converting a chemical signal into an electrical one. Best understood example of a ligand-gated channel. Sometimes called the *nicotinic* acetylcholine receptor to distinguish it from a *muscarinic* receptor, which is a G-protein-linked cell-surface receptor.

Acetylcholine: Neurotransmitter that functions at cholinergic chemical synapses, found both in the brain and in the peripheral nervous system. It is the neurotransmitter at vertebrate neuromuscular junctions.

Acid: In the Brønsted definition, an acid is a chemical species that can donate a proton to another species (a base). In the Lewis definition, an acid is a chemical species that can accept (and share) a pair of electrons from another species. Hydrochloric acid is a Brønsted acid; the proton it donates is a Lewis acid. A neutral Lewis acid and a neutral Lewis base can commonly form a dipolar bond.

Acrosomal process: Long, thin, actin-containing spike produced from the head of certain sperm when they make contact with the egg. Seen in sea urchins and other marine invertebrates whose eggs are surrounded by a thick gelatinous coat.

Acrosome: Region at the head end of a sperm cell that contains a sac of hydrolytic enzymes used to digest the protective coating of the egg.

Actin filament (microfilament): Helical protein filament formed by the polymerization of globular actin molecules. A major constituent of the cytoskeleton of all eucaryotic cells and part of the con- tractile apparatus of skeletal muscle.

Actin: Abundant protein that forms actin filaments in all eucaryotic cells. The monomeric form is sometimes called globular or G-actin; the polymeric form is filamentous or F-actin.

A protein found in many cells. Actin assembles into long polymer rods called microfilaments. These microfilaments form part of the cell's cytoskeleton and they interact with myosin to cause movement in muscles. In muscle contraction, filaments of actin and myosin alternately unlink and chemically link in a sliding action. The energy for this reaction is supplied by adenosine triphosphate (ATP).

Actin-binding protein: Protein that associates with either actin monomers or actin filaments in cells and modifies their properties. Examples include myosin, a-actinin, and profilin.

Action potential: Rapid, transient, self-propagating electrical excitation in the plasma membrane of a cell such as a neuron or muscle cell. Action potentials, or nerve impulses, allow long-distance signaling in the nerveous system.

Activation energy: Extra energy that must be possessed by atoms or molecules in addition to their ground-state energy in order to undergo a particular chemical reaction.

Distinct states correspond to minima of a potential energy surface in a configuration space. In this classical picture, the activation energy for transforming state A into state B is the maximum increase in energy (relative to the ground state of A) encountered on a minimum-energy path from A to B. Energy here refers to potential energy; an analogous definition based on free energy can be constructed. When tunneling is considered, lower energy paths become possible, but an activation energy can be associated with the reaction (at a given temperature) via the relationship between temperature and reaction rate.

Active shield: A defensive system with built-in constraints to limit or prevent its offensive use.

Active site: Region of an enzyme surface to which a substrate molecule binds in order to undergo a catalyzed reaction.

Active transport: Movement of a molecule across a membrane or other barrier driven by energy other than that stored in the concentration gradient or electrochemical gradient of the transported molecule.

Acyclic: Not cyclic.

Acyl group: Functional group derived from a carboxylic acid. (R represents an alkyl group, such as methyl.)

Adenosine triphosphate (ATP): ATP, $C_{10}H_{16}N_5O_{13}P_3$, is an organic compound composed of adenine, the sugar ribose, and three phosphate groups. It serves as the major energy source within the cell to drive a number of biological processes such as photosynthesis, muscle contraction, and the synthesis of proteins.

Adensoine Triphosphate (ATP): A chemical compound that functions as fuel for biomolecular nanotechnology having the formula, $C_{10}H_{16}N_5O_{13}P_3$.

Aerobots (aerobotics): Aerial (flying) robots.

Affinity constant: The reciprocal of the dissociation constant; a measure of the binding energy of a ligand in a receptor.

Affinity: A thermodynamic measurement of the strength of binding between molecules, say between an antibody and antigen. Each antibody/antigen pair has an association constant, Ka, expressed in L/mol.

AFM: An atomic force microscope.

A-kinase (cyclic AMP-dependent protein kinase): Enzyme that phosphorylates target proteins in response to a rise in intracellular cyclic AMP. First identified in skeletal muscle as part of the pathway of regulation of glycogen breakdown in response to adrenaline.

Algorithm: A set of well-defined mathematical rules or operations for solving a problem in a finite number of steps.

Alkane: (1) A saturated, acyclic hydrocarbon structure; usually quite inert. (2) A hydrocarbon containing a double bond; often rather reactive.

Alkyne: A hydrocarbon containing a triple bond; often rather reactive.

AM 1: The air mass 1 spectrum of a light source is equivalent to that of sunlight at the earth's surface when the sun is at zenith.

Amide: A molecule containing an amine bonded to a carboxyl group; the resulting bond has substantial double-bond character. Also termed a peptide; amide bonds link amino acids in proteins.

Amine: A molecule containing N with a single bond to C and two other single bonds to H or C (but not an amide); the amine group or moiety.

Amino acid: A molecule containing both an amine and a carboxylic acid group; in the 20 genetically encoded amino acids in biology, both groups are bound to the same C. Amino acids joined by amide bonds form peptides and proteins; these do not contain amino acids as such, and are instead said to contain *amino acid residues*. Organic molecules that are the building blocks of proteins. There are some two hundred known amino acids, of which twenty are used extensively in living organisms.

Ampere (amp) [A]: Measure of electric current: 1A = 1 coulomb/second.

Amperometric Sensor: Amperometric sensors involve a heterogeneous electron transfer as a result of an oxidation/reduction of an electro-active species at a sensing electrode surface. A current is measured at a certain imposed voltage of the sensing electrode with respect to the reference electrode. Analytical information is obtained from the current-concentration relationship at that given applied potential.

Analyte: A chemical species targeted for qualitative or quantitative analysis.

Angstrom [Å]: Measure of length: 1 Å= $1x10^{-10}$ m.

Angstrom: A unit of length equal to 0.1 nanometer = 10^{10} meters.

Anion: A negatively charged ion.

Anisotropic: Exhibiting different values of a property in different crystallographic directions.

Anneal: Heat process used to remove stress, crystallize or render deposited material more uniform.

Anode: The electrode in an electrochemical cell or galvanic couple that experiences oxidation, or gives up electrons.

Antioxidants: Chemicals that protect against oxidation, which causes rancidity in fats and damage to DNA.

Armored: The Molecular Plate™ Advantage Evident's Molecular Plate™ covering for InGaP EviDots is lattice matched - with a matched pattern of atoms - to improve its molecular bonding to the core. This results in a quantum dot with high brightness and increased stability for longer lasting fluorescence.

Aromatic: A term used to describe cyclic pi-bonded structures of special stability.

Arrhenius equation: The equation representing the rate constant as $k = Ae^{Ea/RT}$ where A represents the product of the collision frequency and a steric factor, and $e^{Ea/RT}$ is fraction of collisions with sufficient energy to produce a reaction.

Artificial Intelligence (AI): A field of research that aims to understand and build intelligent machines; this term may also refer to an intelligent machine itself.

ASIC: Application-specific integrated circuit; an IC designed for a custom requirement.

Assembler: A chemical device that given certain atomic or molecular inputs (starting materials) can output a specific molecular structure or aggregation.

A general-purpose device for molecular manufacturing capable of guiding chemical reactions by positioning molecules.

A molecular machine that can be programmed to build virtually any molecular structure or device from simpler chemical building blocks. Analogous to a computer-driven machine shop.

A nano-robotic device controlled by an onboard computer that can use available chemicals to manufacture nanoscale products. It has been proposed that advanced designs could communicate, cooperate, and maneuver to build macroscale products. Assemblers are much more complex, and probably less efficient, than fabricators.

In recent popular usage, any nanomachine, usually assumed to offer magical, universal capabilities in an atom-sized package. In the author's usage, any programmable nanomechanical system able to perform a wide range of mechanosynthetic operations. See molecular manipulator, molecular mill.

At the Core: A Non-Heavy Metal Quantum Dot. Research at Evident, which is focused on advancing product development, led to the first commercially available nanocrystals that comprise a III-V semiconductor material: Indium, Gallium and Phosphide core quantum dots, with a unique molecular metallic platting compound to form the shell or InGaP/ZnS EviDots-MP™.

Atom: The smallest particle of a chemical element (about three ten-billionths of a meter in diameter). Atoms are the building blocks of molecules and solid objects; they consist of a cloud of electrons surrounding a dense nucleus a hundred thousand times smaller than the atom itself. Nanomachines will work with atoms, not nuclei. (2) The smallest unit of a chemical element, about a third of a nanometer in diameter. Atoms make up molecules and solid objects.

Atomic Force Microscope (AFM): An instrument able to image surfaces to molecular accuracy by mechanically probing their surface contours. A kind of proximal probe. A device in which the deflection of a sharp stylus mounted on a soft spring is monitored as the stylus is moved across a surface.

If the deflection is kept constant by moving the surface up and down by measured increments, the result (under favorable conditions) is an atomic-resolution topographic map of the surface. Also termed a scanning force microscope. See How AFM Works, What is an Atomic Force Microscope? and Window on a Small World

Atomic Force Microscopy (AFM): An instrument that uses a sharp tip to probe a surface. By moving the tip around, an image of the surface can be made. Height differences as small as 10^{-11} m can be measured this way. (2) A device in which the deflection of a sharp stylus mounted on a soft spring is monitored as the stylus is moved across a surface. If the deflection is kept constant by moving the surface up and down by measured increments, the result (under favorable conditions) is an atomic-resolution topographic map of the surface. Also termed a scanning force microscope. (3) A type of scanning probe microscopy that maps the topography of an interface by scanning a force sensor over the interface.

Atomic Manipulation: Manipulating atoms, typically with the tip of an STM.

Atomic mass unit (a.m.u.): A unit of mass used to express relative atomic masses. It is equal to 1/12 of the mass of an atom of the isotope carbon-12 and is equal to 1.66033×10^{-27} kg

Atomic number (also proton number Z): The number of protons within the atomic nucleus of a chemical element.

Atomic weight: The weighted average mass of the atoms in a naturally occurring element.

Atomistic Simultations: Atomic motion computer simulations of macromolecular systems are increasingly becoming an essential part of materials science and nanotechnology. Recent advances in supercomputer simulation techniques provide the necessary tools for performing computations on nanoscale objects containing as many as 300,000 atoms and on materials simulated with 1,000,000 atoms. This new capability will allow computer simulation of mechanical devices or molecular machines using nanometer size components. [The Center for Computational Sciences at the ORNL]

Atoms: Atoms are the smallest particle of matter that can be classified as being a specific element. Atoms are made up of electrons, protons and neutrons. The positively charged protons and the neutrally charged neutrons reside in the nucleus at the center of the atom. A atomic nucleus is very small; it has a radius of $10^{-15} - 10^{-14}$ meters. Nearly all of the mass of an atom is concentrated in the nucleus. The negatively charged electrons are found in a roughly spherical region with a radius of about 10^{-10} meters.

The number protons determines which element an atom is. For instance, carbon atoms always have 6 protons. If an atom has more or less electrons than protons then it is charged and is called an ion. Two atoms with the same number of protons but different numbers of neutrons are two different isotopes of the same element.

ATP (Adenosine Triphosphate): A chemical compound that functions as fuel for biomolecular nanotechnology having the formula, $C_{10}H_{16}N_5O_{13}P_3$.

ATPase: A protein complex responsible for converting electical potential energy (from a nanoscale batery) into ATP, the molecule used for immediate fuel by virtually all living systems.

Austenite: Face-centered cubic iron; also iron and steel alloys that have the FCC crystal structure.

Autogenous control: In medical nanorobotics, the conscious control of in vivo nanorobotic systems by the human user or patient; in biochemistry, the action of a gene product that either inhibits (negative autogenous control) or activates (positive autogenous control) expression of the gene coding for it.

Automated Chemistry: The use of an automated machine (e.g. computer driven robot or fluidic devices) to carry out chemical reactions, purification and molecular assembly.

Automated Engineering: Engineering design done by a computer system, generating detailed designs from broad specifications with little or no human help.

The use of computers to perform engineering design, ultimately generating detailed designs from broad specifications with little or no human help. Automated engineering is a specialized form of artificial intelligence.

Automated manufacturing: As used here, nanotechnology-based manufacturing requiring little human labor.

Autoproductivity: The ability of a system, under external control, to automatically produce an identical copy of itself.

Avogadro's number: The number of atoms in exactly 12 grams of pure ^{12}C, equal to 6.022×10^{23}.

B

B lymphocyte (B cell): Type of lymphocyte that makes antibodies.

Bacteria: One-celled living organisms, typically about one micron in diameter. Bacteria are among the oldest, simplest, and smallest types of cells. Single-celled microorganisms, about one micrometer (one thousand nanometers) across.

Bacteriophage (phage): Any virus that infects bacteria. They were the first organisms used for the study of molecular genetics and are now widely used as cloning vectors. (From Greek *phagein*, to eat.)

Bacteriorhodopsin: Pigmented protein found in the plasma membrane of a salt-loving bacterium, *Halobacterium halobium*; it pumps protons out of the cell in response to light.

Bacterium (plural bacteria): Common name for any member of the diverse group of procaryotic organisms. Most are single cells, but multicellular forms also exist.

Ballistic Magnetoresistance (BMR): Ballistic Magnetoresistance is yet another way in which spin orientation, encoding information on a storage medium such as a hard drive, can modify electrical resistance in a nearby circuit, thereby accomplishing the sensing of that orientation. [Physics News]

Band gap energy (E_g): For semiconductors and insulators, the energies that lie between the valence and conduction bands; for intrinsic materials, electrons are forbidden to have energies within this range.

Bandwidth: The range of frequencies over which the measurement system can operate within a specified error range.

Banoelectronics: Nanoelectronics in Japan and Korea tends to focus on next generation semiconductor devices and *single electron devices (SED).* [Asia Pacific Nanotechnology Initiatives: Update Asian Technology Information Program, Japan April 10, 2001]

Banopore: Nanopore technology is an elegant concept. A membrane with very small channels (a few nanometers in diameter), called nanopores, separates two solutions. When a voltage is applied across the membrane, charged biomolecules migrate through the pores in a controlled manner.... As each nucleotide passes through the nanopore, an electronic signature is produced that can be used to characterize it.

The nanopore's size is such that it permits only a single nucleic acid strand to pass through it at any one time. Therefore the technology can be used to sequentially measure a biopolymer's properties along its length.

Research in the field of *computational biology* will play a significant role in the development of nanopore technology. It will be used to help optimize experimental strategies and designs, and to create the mathematical methods needed to interpret the data generated.

The potential capabilities of nanopore technology are very broad. Nanopore technology can distinguish between and count a variety of different molecules in a complex mixture. For example, nanopores could discriminate between hybridized or unhybridized unknown RNA and DNA molecules that differ by a single nucleotide only. Compared to existing techniques, nanopore technology is expected to provide direct characterization of individual nucleic acid and protein molecules directly derived from biological samples, thereby making it applicable to a wide set of analyses. Because nanopore technology is in the very early stages of its development, its advantages are not fully characterized.

Banopositioning: The means of controlling motion on the nanometre scale—is now a key enabling technology in high- tech fields such as semiconductor test and measurement, photonics alignment, scanning microscopy and microlithography.

Banoprism: Scientists at Northwestern University have created a nanoparticle with a new shape that could be a useful tool in the race to detect biological threats. The nanoprism, which resembles a tiny Dorito, exhibits unusual optical properties that could be used to improve biodetectors, allowing them to test for a far greater number of biological warfare agents or diseases at one time.

Banoscale: 1 to 100 billionths of a meter. At the nanoscale, physics, chemistry, biology, materials science, and engineering converge toward the same principles and tools. The nanoscale is not just another step toward *miniaturization*, but a qualitatively new scale. The new behavior is dominated by quantum mechanics, material confinement in small structures, large interfacial volume fraction, and other unique properties, phenomena and processes. Many current theories of matter at the microscale have critical lengths of nanometer dimensions. These theories will be inadequate to describe the new phenomena at the nanoscale. ... Innovative nanoscale properties and functions will be achieved through the control of matter at its building blocks: atom- by- atom, molecule- by-molecule, and nanostructure- by-nanostructure.

Nanotechnology will include the integration of these nanoscale structures into larger material components, systems, and architectures. However, within these larger scale systems the control and construction will remain at the nanoscale. [National Science Foundation, Societal Implications of Nanoscience and Nanotechnology, Report of Sept 28- 29 2000 workshop, Mar. 2001] http://itri.loyola.edu/nano/societalimpact/nanosi.pdf

Banoscience: The study of phenomena and manipulation of materials at atomic, molecular and macromolecular scales, where properties differ significantly from those at a larger scale.

Nanoscience is primarily the extension of existing sciences into the realms of the extremely small (*nanomaterials, nanochemistry, nanobio, nanophysics*, etc.) while *nanoengineering* represents the extension of the engineering fields into the nano- scale realm (*nanofabrication, nanodevices*, etc.).

The exponential growth of nanoscience is largely due to the development of new instruments and related techniques that are used to "routinely" probe and manipulate material at the atomic and molecular level. *Scanning probe microscopies,* analytical electron- beam techniques, epitaxial growth facilities, and *synchrotron* radiation sources are all opening huge opportunities.

Banoshells: Procedures that target cancer cells while leaving normal cells untouched, patients controlling the release of medicine in their bodies with an infrared light, and medical test results produced in seconds rather than days – these are three new medical technologies currently being tested by nanotechnology researchers at Rice University. ... The research focuses on nanoshells, a new type of nanoparticle invented by Naomi Halas, Rice professor of electrical and computer engineering and chemistry. Nanoshells are layered nanoparticles whose ability to manipulate light and color can be designed into the nanoparticle by varying the thickness of the nanoparticle's layers. [Rice Univ. Media Release "New nanotechnology has medical applications" April 12, 2001] http://www.rice.edu/projects/reno/Newsrel/2001/20010412_nanoshell.shtml

Baronatation: In medical nanorobotics, locomotion through a frozen fluid by applying mechanical pressure along the path travelled to induce melting ahead, followed by regelation (refreezing) behind.

Barrier height: Roughly synonymous with activation energy.

Basal body: Short cylindrical array of microtubules plus their associated proteins found at the base of a eucaryotic cell cilium or flagellum. Serves as a nucleation site for the growth of the axoneme. Closely similar in structure to a centriole.

Basal lamina (plural basal laminae): Thin mat of extracellular matrix that separates epithelial sheets, and many types of cells such as muscle cells or fat cells, from connective tissue. Sometimes called a basement membrane.

Basal: Situated near the base. The basal surface of a cell is opposite the apical surface.

Base pair: Two nucleotides in an RNA or a DNA molecule that are paired by hydrogen bonds - for example, G with C and A with T or U.

Base: In the Brønsted definition, a base is a chemical species that can accept a proton from another species. In the Lewis definition, a base is a chemical species that can donate (and share) a pair of electrons with another species. Molecule (usually containing nitrogen) that accepts a proton in solution. Often used to refer to the purines and pyrimidines in DNA and RNA.

Basic: Having the properties of a base.

Basophil: White blood cell that releases histamine in an inflammatory response. Closely related to a mast cell.

BAW: Bulk acoustic wave.

Bbionanotechnology: Includes molecular motors, biomaterials, single molecule manipulation technologies, biochip technologies, etc. [Asia Pacific Nanotechnology Initiatives: Update Asian Technology Information Program, Japan April 10, 2001] www.atip.or.jp/public/atip.reports.01/atip01.018.pdf

Bearing: A mechanical device that permits the motion of a component (ideally, with minimal resistance) in one or more degrees of freedom while resisting motion (ideally, with a stiff restoring force) in all other degrees of freedom.

BECON Bioengineering Consortium: The focus of bioengineering activities at the NIH. The Consortium consists of senior- level representatives from all of the NIH institutes, centers, and divisions plus representatives of other Federal agencies concerned with biomedical research and development. BECON is administered by the National Institute of Biomedical Imaging and Bioengineering (NIBIB).

Bilayer lipid membrane (BLM): The structure found in most biological membranes, in which two layers of lipid molecules are so arranged that their hydrophobic parts interpenetrate, whereas their hydrophilic parts form the two surfaces of the bilayer.

Binary: Numbering system based on powers of 2 using only the digits 0 and 1, called "bits".

Binding energy: The reduction in the free energy of a system that occurs when a ligand binds to a receptor. Generally used to describe the total energy required to remove something, or to take a system apart into its constituent particles—for example, to separate two atoms from one another, or to separate an atom into electrons and nuclei.

Binding site: The active region of a receptor; any site at which a chemical species of interest tends to bind.

Binding: The process by which a molecule (or ligand) becomes bound, that is, confined in position (and often orientation) with respect to a receptor. Confinement occurs because structural features of the receptor create a potential well for the ligand; van der Waals and electrostatic interactions commonly contribute.

Bio-assemblies or Biomolecular Assemblies: Containing several protein units, DNA loops, lipids, various ligands, etc.

Biochauvinism: The prejudice that biological systems have an intrinsic superiority that will always give them a monopoly on self-reproduction and intelligence.

Biocompatible coated materials: Biocompatible materials usually used in dental and bone implants that enhance biologic fixation, thereby increasing the bond strength between the coated material and bone, and minimize possible biological effects that may result from the implant itself. MeSH, 1999

Biocompatible materials: Synthetic or natural materials, other than drugs, that are used to replace or repair any body tissue or bodily function. MeSH, 1973

Biodefense: The need for improved national defenses against biological attacks are urgently needed. A real-time reporting system needs to be developed, whereby officials can be informed about an emerging threat before an agent has a chance to affect thousands of people. The development of integrated systems for detecting and monitoring biological agents is instrumental to this goal. The development of vaccines and novel detection platforms will enable preparedness, while also benefiting human health.

Biodynotics Biologically Inspired Multifunctional Dynamic Robotics: A multidisciplinary, multi-pronged approach with far reaching impact on robotic capabilities for national security applications. Biologically Inspired Multifunctional Dynamic Robotics: BIODYNOTICS will explore the following areas: DYNAMIC MOBILITY, BEHAVIOR,

INTEGRATION Biological Sciences, Defense Sciences Office, DARPA.

Bioengineering: Is rooted in physics, mathematics, chemistry, biology, and the life sciences. It is the application of a systematic, quantitative, and integrative way of thinking about and approaching the solutions of problems important to biology, medical research, clinical proactive, and population studies. The NIH Bioengineering Consortium agreed on the following definition for bioengineering research on biology, medicine, behavior, or health recognizing that no definition could completely eliminate overlap with other research disciplines or preclude variations in interpretation by different individuals and organizations.

Integrates physical, chemical, or mathematical sciences and engineering principles for the study of biology, medicine, behavior, or health. It advances fundamental concepts, creates knowledge for the molecular to the organ systems levels, and develops innovative biologics, materials, processes, implants, devices, and informatics approaches for the prevention, diagnosis, and treatment of disease, for patient rehabilitation, and for improving health.

Biofabrication: Includes nanoparticle delivery systems, biomaterials, tissue engineering, implants and prostheses. [BECON, Nanoscience and Nanotechnology:

Using biological processes to synthesize and manufacture chemicals and materials of high value to the Department of Defense. Biological processes are characterized by: low energy barriers (~10Kcal, high temperatures/pressures not required); high reaction-, regio- and stereo- specificity (protection/

deprotection wastes and catalyst poisoning avoided); spatio-temporal control of materials synthesis of defined composition and size with angstrom- level precision; and local control of the dielectric environment (largely eliminating the need for toxic solvents). Potential target materials and chemicals include composites with enhanced mechanical properties, ultra-low k dielectrics and thermoelectrics, optoelectronic materials, photonic devices (waveguides and logic elements), electronic materials (e.g., GaN, InGaN, AlGaN), elastomers, energetic materials, and adhesives. Biofabrication, DARPA.

Biomaterials: Synthetic or natural materials that can replace or augment tissues, organs or body functions.

Biomechanics: Mechanical structures of living organisms (especially muscles and bones).

Biomedical Nanotechnology: (1) the comprehensive monitoring, control, construction, repair, defense, and improvement of all human biological systems, working from the molecular level, using engineered nanodevices and nanostructures; (2) the science and technology of diagnosing, treating, and preventing disease and traumatic injury, of relieving pain, and of preserving and improving human health, using molecular tools and molecular knowledge of the human body; (3) the employment of molecular machine systems to address medical problems, using molecular knowledge to maintain and improve human health at the molecular scale. (4) A rapidly expanding field that includes many potential technologies and approaches. The key to this definition is that phenomena and materials at the nanometer scale are known to have properties that are uniquely attributable to that scale length. Nanomedicine could similarly be defined [as nanotechnology] as the design, synthesis, or application of materials, devices, or technologies in the nanometer-scale for the basic understanding, diagnosis, and / or treatment of disease. Canadian Institute of Health Research, Regenerative Medicine and Nanomedicine, RFA, 2003. The monitoring, repair, construction and control of human biological systems at the molecular level, using engineered nanodevices and nanostructures. (5) the comprehensive monitoring, control, construction, repair, defense, and improvement of all human biological systems, working from the molecular level, using engineered nanodevices and nanostructures; (6) the science and technology of diagnosing, treating, and preventing disease and traumatic injury, of relieving pain, and of preserving and improving human health, using molecular tools and molecular knowledge of the human body; (7) the employment of molecular machine systems to address medical problems, using molecular knowledge to maintain and improve human health at the molecular scale.

The use of molecular-scale devices (nanotechnology) to repair damage and boost the immune system. Vapour grown carbon fibres are obtained as shown in this SEM image. The diameters of these fibres can vary from 100 nm to 500 nm. As part of the National Institutes of Health (NIH) Roadmap for Medical Research [nihroadmap.nih.gov], the NIH [nih.gov] will establish a handful of nanomedicine centers. These centers will be staffed by a highly interdisciplinary scientific crew including biologists, physicians, mathematicians, engineers and computer scientists. Research conducted over the first few years will be spent gathering extensive information about how molecular machines are built. A key activity during this time will be the development of a new kind of vocabulary, or lexicon, to define biological parts and processes in engineering terms.

Once researchers have completely catalogued the interactions between and within molecules, they can begin to look for patterns and a higher order of connectedness than is possible to identify with current experimental methods. Mapping these networks and understanding how they change over time will be a crucial step toward helping scientists understand nature's rules of biological design. Understanding these rules will, in many years' time, enable researchers to use this information to address biological issues in unhealthy cells.

Biomedical polymers: More and more therapeutic problems are relevant to the use of polymer- based therapeutic aids for a limited period of time, namely the healing time related to the outstanding capacity of living systems to self- repair ... After healing the remaining prosthetic materials or devices become foreign residues or wastes that have to be eliminated from the body.

Nowadays, biocompatible polymers that can degrade in the body are developed. The degradation and the elimination of degradation by-products depend on rather complex phenomena that are presently reflected inconsistently by terms issued from the tradition because each domain has developed its own terminology almost independently. This is a source of misunderstandings, confusions and misperceptions among scientists, surgeons, pharmacists, journalists and politicians, the situation being increased by the introduction of degradable polymers in plastic waste management and environmental protection. Therefore, it is urgent to reflect the various phenomena by specific terms, harmonize and enforce their use by the people active in the biomedical, pharmacological and environmental fields, and, last but not least by the publishing media and journalists. IUPAC, Terminology for biomedical (therapeutic) polymers.

BioMEMS Biological MicroElectro Mechanical Systems: Highlights the technical advances in the field that are leading a revolution in medicine, and creating a new generation of analytical devices for medical diagnostics. The meeting will encompass technology developments in micro & nano drug delivery, interface of nanotech and tissue engineering, microfluidics, and miniaturized total analysis systems (microTAS), biosensors, innovations in mass spec, and nanoscale imaging *BioMEMS and Nanotech World*, Aug. 16- 17, 2004, Washington DC

BioMEMS: MEMS used in medicine, that use microchips.

Biomimetic Chemistry: Knowledge of biochemistry, analytical chemistry, polymer science, and biomimetic chemistry is linked and applied to research in designing new molecules, molecular assemblies, and macromolecules having biomimetic functions. These new bio-related materials of high performance, including, for example, enzyme models, synthetic cell membranes, and biodegradable polymers, are prepared, tested, and constantly improved in this division for industrial scale production. [DCBE]

Biomimetic materials: Materials fabricated by BIOMIMETICS techniques, i.e., based on natural processes found in biological systems. [MeSH 2003] (2) Materials that imitate, copy, or learn from nature.

Biomimetic: Imitating, copying, or learning from biological systems.

Biomimetic: Imitating, copying, or learning from nature. Nanotechnology already exists in nature; thus, nanoscientists have a wide variety of components and tricks already available.

Biomimetics: An interdisciplinary field in materials science, ENGINEERING, and BIOLOGY, studying the use of biological principles for synthesis or fabrication of BIOMIMETIC MATERIALS.

There is a need to develop the next generation of restorative materials and medical implants. New avenues of scientific inquiry may enable the development of biomaterials that are safe, reliable, "smart", long-lasting, and perform ideally in their respective biological environments.... Over the last few years biomimetics and tissue engineering have emerged as a new vision in the field of tissue and organ repair and restoration. Biomimetics and tissue engineering are interdisciplinary fields that combine information from the study of biological structures and their functions with physics, mathematics, chemistry and engineering for the generation of new materials, tissues and organs. These approaches can offer new ways of:

(a) developing biological solutions for future design and synthesis of composite materials such as bone, cartilage, tendon, ligament, skin, dentin, enamel, cementum and periodontal ligament;

(b) replacing and assembling functional tissues and organs; and

(c) evaluating medical and dental implants.

In the area of craniofacial, oral and dental principles from biomimetics and tissue engineering are applied to developing dental and facial implants, new polymers for guided tissue regeneration used in treating periodontal disease and bone and connective tissue defects, coral- based hydroxyapatite replicas for reconstruction of alveolar ridges and other osseous defects, temporomandibular joint (TMJ) and other joint prostheses, formation of bone matrix substitutes, and artificial replicas of bone, skin, and mucosa.

The term biomimetics was coined in 1972 in the context of artificial enzymes; it might be defined broadly as "the abstraction of good design from nature". It is a fact of everyday life that nature has managed to built materials and 'devices' with breathtaking functionality, heterogeneity and stability by using a comparatively limited number of building blocks (the whole range of synthetic materials is restricted to man- made engineering). The basic concepts of nature are often simple; it is the way in which building blocks and materials are arranged that results in functionality. Among the most simple and abundant themes of nature is self-assembly: lipids assemble in sheets to form cell membranes, proteins assemble into functional enzymes, cellular 'sensors', fibers, or virus coats, and DNA assembles in double strands to provide the very basis for live: replication. Biomimetics Study of the structure and function of biological substances to make artificial products that mimic the natural ones.

BIOML Biopolymer Markup Language: Designed by the BIOML core team at Proteometrics, LLC and Proteometrics Canada Ltd. It is to be used "for the annotation of biopolymer sequence information. BIOML allows the full specification of all experimental information known about molecular entities composed of biopolymers, for example, proteins and genes."

Biomolecular materials: An emerging discipline, materials whose properties are abstracted from biology. They share many of the characteristics of biological materials but are not necessarily of biological origin. For example, they may be inorganic materials that are organized or processed in a biomimetic fashion. A key feature of biological and biomolecular materials is their ability to undergo *self- assembly*.

Biomolecular Nanotechnology: Nanotechnology existing in living systems and resulting from our ability to use biomolecules as components for molecular nanotechnology.

Biomotors: Driven by energy sources such as *adenosine triphosphate (ATP)* for chemical transduction and other processes. These biomotors are considered to be biomolecular and are discussed in the body of this report, but strictly speaking they do not conform to the panel's definition of self- assembly. [Biomolecular self- assembling materials, National Academy of Sciences 1996]

BioNEMS: biofunctionalized nanoelectromechanical systems.

Biopolymeroptoelectromechanical Systems [BioPOEMS]: Combining optics and microelectromechanical systems, and used in biological applications.

Biopolymers: Macromolecules (including proteins, nucleic acids and polysaccharides) formed by living organisms. [IUPAC Compendium]

Biorobotics: Our research focuses on the role of sensing and mechanical design in motor control, in both robots and humans. This work draws upon diverse disciplines, including biomechanics, systems analysis, and neurophysiology. The main approach is experimental, although analysis and simulation play important parts. In conjunction with industrial partners, we are developing applications of this research in biomedical instrumentation, teleoperated robots, and intelligent sensors.

Biosensor: The term "biosensor" is a general designation that denotes either a sensor to detect a biological substance or a sensor which incorporates the use of biological molecules such as antibodies or enzymes. Biosensors are a subcategory of chemical sensors.

Biostasis: A condition in which an organism's cell and tissue structure are preserved, allowing later restoration by cell repair machines. Applicable to cryonics. See also "ischemic coma," "ametabolic coma," "biostatic coma," and "in suspension." [Brian Wowk]

Biovorous: From "biovore;" an organism capable of converting biological material into energy for sustenance.

Bipolar-junction transistor: Transistor with n-type and p-type semiconductors having base-emitter and collector-base junctions.

Bit: Numbering system based on powers of 2 using only the digits 0 and 1, called "bits".

Blue Goo: Opposite of Grey goo. Benificial tech, or "police" nanobots.

Body-centered cubic (BCC): A common crystal structure found in some elemental metals. Within the cubic unit cell, atoms are located at corner and cell center positions.

Bogosity Filter: A mechanism for detecting bogus ideas and propositions.

Bond: Two atoms are said to be bonded when the energy required to separate them is substantially larger than the van der Waals attraction energy. Ionic bonds result from the electrostatic attraction between ions; covalent and metallic bonds result from the sharing of electrons among atoms; hydrogen bonds are weaker and result from dipole interactions and limited electron sharing. When used without modification, "bond" usually refers to a covalent bond.

Bone substitutes: Synthetic or natural materials for the replacement of bones or bone tissue. They include hard tissue replacement polymers, natural coral, hydroxyapatite, beta- tricalcium phosphate, and various other biomaterials. The bone substitutes as inert material's can be incorporated into surrounding tissue or gradually replaced by original tissue. [MeSH, 1995]

Born-Oppenheimer Approximation: Permits the use of classical mechanics in modeling and thinking about molecular and atomic motions. Needless to say, this greatly simplifies the conceptual framework required for thinking about molecular machines. [RCM] Once used as an argument on why MNT could not work. Since refuted: See That's impossible! How good scientists reach bad conclusions

Bose-Einstein Condensates [BEC's]: Bose-Einstein Condensates "...aren't like the solids, liquids and gases that we learned about in school. They are not vaporous, not hard, not fluid. Indeed, there are no ordinary words to describe them because they come from another world — the world of quantum mechanics." [See A New Form of Matter]

Bottom Up: Building larger objects from smaller building blocks. Nanotechnology seeks to use atoms and molecules as those building blocks. The advantage of bottom-up design is that the covalent bonds holding together a single molecule are far stronger than the weak. Mostly done by chemists, attempting to create structure by connecting molecules.

Bottom-up fabrication: Building a small structure by assembling it out of atoms or small molecules.

Bottom-up nanotechnology: Mostly chemists attempting to create structure by connecting molecules. [Noah Robischon "Nanotechnology and the battle to build smaller" Discovery Channel 1998]

Brazing: A metal joining technique that uses a molten filler metal alloy having a melting temperature greater than about 425°C (800°F).

Breakdown: Failure of a material resulting from an electrical overload. The resulting damage may be in the form of thermal damage (melting or burning) or electrical damage (loss of polarization in piezoelectric materials). The failing of a physical principle or unit of measurement as the scale is changed.

Brownian Assembly: Brownian motion in a fluid brings molecules together in various position and orientations. If molecules have suitable complementary surfaces, they can bind, assembling to form a specific structure. Brownian assembly is a less paradoxical name for self-assembly (how can a structure assemble itself, or do anything, when it does not yet exist?).

Brownian Motion: Motion of a particle in a fluid owing to thermal agitation, observed in 1827 by Robert Brown. (Originally thought to be caused by vital force, Brownian motion in fact plays a vital role in the assembly and activity of the molecular structures of life).

Buckminsterfullerene: A fullerene with the chemical formula (C_{60}). See Fullerenes. A broad term covering the variety of buckyballs and carbon nanotubes that exist. Named after the architect Buckminster Fuller, who is famous for the geodesic dome, which buckyballs resemble.

Bucky Balls [AKA - C60 molecules & buckminsterfullerene]: Molecules made up of 60 carbon atoms arranged in a series of interlocking hexagonal shapes, forming a structure similar to a soccer ball. See our Nanotubes and Buckyball page.

Buckyball or Fullerene or C60: One of three known pure forms of carbon (graphite and diamond being the other two) that takes a spherical shape with a hollow interior. Buckyballs, named because they resemble the geodesic domes built by architect Buckminster Fuller, were discovered in 1985 among the byproducts of laser vaporization of graphite in which the carbon atoms are arranged in sheets.

Though C60, referring to the number of carbon atoms that make up one sphere, is the most common fullerene, researchers have found stable, spherical carbon structures containing 70 atoms (C70), 120

(C120), 180 (C180), and others. Robert F. Curl Jr. and Richard E. Smalley, both of Rice University in Houston, Texas and Harold W. Kroto of the University of Sussex in England, won the 1996 Nobel Prize for Chemistry for their discovery of buckminsterfullerene, the scientific name for buckyballs.

One of three known pure forms of carbon (graphite and diamond being the other two) that takes a spherical shape with a hollow interior. Buckyballs, named because they resemble the geodesic domes built by architect Buckminster Fuller, were discovered in 1985 among the byproducts of laser vaporization of graphite in which the carbon atoms are arranged in sheets. Though C60, referring to the number of carbon atoms that make up one sphere, is the most common fullerene, researchers have found stable, spherical carbon structures containing 70 atoms (C70), 120 (C120), 180 (C180), and others.

Buckyball: Round (convex) fullerenes ranging in size from 20 to over 500 carbon atoms.

Bulk Technology: Technology based on the manipulation of atoms and molecules in bulk, rather than individually; most present technology falls in this category. (2) Technology in which atoms and molecular are manipulated in bulk, rather than individually.

Bus: Transmission medium for electrical or optical signals that perform a particular function, such as computer control.

Bush Robot: A concept for robots of ultimate dexterity, they utilize fractal branching to create ever-shrinking "branches," eventually ending in nanoscale "fingers." Developed by Hans Moravec. See Fractal branching ultra-dexterous robots

Byte: A group of eight bits that can represent any of 2^8 = 256 different entities.

C

C_{60}: A molecule consisting of 60 carbon atoms arranged in the pattern found on a soccer ball. C_{60} molecules are also known as Buckministerfullerine and as Buckyballs. Nanotech Now's nanotube and buckyball page

Ca2+-release channel: Ion channel in the membrane of the endoplasmic reticulum and sarcoplasmic reticulum (in muscle cells) that releases Ca2+ into the cytosol when activated.

CAD: Computer-aided design.

Cadherin: Member of a family of proteins that mediate Ca2+-dependent cell-cell adhesion in animal tissues.

Caged compound: Organic molecule designed to change into an active form when irradiated with light of a specific wavelength. An example is caged ATP.

Calibration: A process of adapting a sensor output to a known physical quantity to improve sensor output accuracy. Ubiquitous Ca^{2+}-binding protein whose binding to other proteins is governed by changes in intracellular Ca^{2+} concentration. Its binding modifies the activity of many target enzymes and membrane transport proteins.

Calorie: Unit of heat. One calorie (small "c") is the amount of heat needed to raise the temperature of 1 gram of water by 1 °C. A kilocalorie is the unit used to describe the energy content of foods.

Calvin cycle (Calvin-Benson cycle): Major metabolic pathway by which CO2 is fixed during photosynthesis.

Cam: A component that translates or rotates to move a contoured surface past a follower; the contours impose a sequence of motions (potentially complex) on the follower.

cAMP: Enzyme that phosphorylates target proteins in response to a rise in intracellular cyclic AMP. First identified in skeletal muscle as part of the pathway of regulation of glycogen breakdown in response to adrenaline. See cyclic AMP

Cantilever: A lever beam held down at one end, with some support near the middle and which supports a load on the other end. Diving boards and drawbridges are cantilevers.

CAP (catabolite gene activator protein): Gene regulatory protein in procaryotes that, when glucose is absent, activates genes responsible for the breakdown of alternative carbon sources.

Capacitance (C): The charge-storing ability of a capacitor, defined as the magnitude of charge stored on either plate divided by the applied voltage. A 1-F capacitor charged to 1 V contains C of charge (see also capacitor) and 1 C is an amount of charge equal to that of about 6.24×10^{18} electrons.

Capacitor: Energy storage circuit element having two conductors separated by an insulator.

Capillaries: Microscopic blood vessels that carry oxygenated blond to tissues.

Capsid: Protein coat of a virus, formed by the self-assembly of one or more protein subunits into a geometrically regular structure.

Carbanion: A highly reactive anionic chemical species with an even number of electrons and an unshared pair of electrons on a tetravalent carbon atom.

Carbene: A highly reactive chemical species containing an electrically neutral, divalent carbon atom with two nonbonding valence electrons; a prototype is CH_2.

Carbon nanofoam: A new form of carbon: a spongy solid that is extremely lightweight and, unusually, attracted to magnets... John Giapintzakis of the University of Crete has used an electron microscope to study the structure of the nanofoam. He says it is the fifth form of carbon known after graphite, diamond and two recently discovered types: hollow spheres, known as buckminsterfullerenes or buckyballs, and nanotubes. Electrically conductive carbon nanofoams are a new material with many of the properties of traditional aerogel material. These materials are available in the form of monoliths, granules, powders and papers. They are synthetic, lightweight foams in which the solid matrix and pore spaces have nanometer- scale dimensions. Prepared by sol- gel methods, nanofoams typically have low density, continuous porosity, high surface area, and fine cell/ pore sizes. Carbon nanotubes are cylindrical structures made only from carbon atoms that are about 1 nm in diameter and 1-100 microns in length. In a graphite sheet, carbon atoms arrange themselves in a two-dimensional hexagonal lattice. Carbon nanotubes can be though of as a strip of graphite sheet that is rolled up to form a cylinder.

There are many different ways to cut up a piece of graphite and roll it up to form a tube. The tubes can have different diameters and different chiralities. The chirality is the twist of the rows of atoms along the length of the tube. Sometimes the atom rows are parallel to the axis of the tube and sometimes the rows form a helix that winds along the tube. When one or more tubes grow inside another carbon nanotube, it is called a multiwalled nanotube. Carbon naotubes are very strong; they are 5 times as strong as steel for the same wieght.

The electrical properties of carbon nanotubes depend on their diameter and their chirality. Some tubes are metallic and some are semiconductors. Research on carbon nanotubes at TU Delft, Nanotech Now's nanotube and buckyball page. A form of carbon related to fullerenes, except that the carbon atoms form extended hollow tubes instead of closed, hollow spheres.

Carbon nanotubes can also form as a series of nested, concentric tubes. Carbon nanotubes can be used as nanometer-scale syringe needles for injecting molecules into cells and as nanoscale probes for making fine-scale measurements. Carbon nanotubes can be filled and capped, forming nanoscale test tubes or potential drug delivery devices. Carbon nanotubes can also be "doped," or modified with small amounts of other elements, giving them electrical properties that include fully insulating, semiconducting, and fully conducting.

Carbon Nanotubes: A form of carbon related to fullerenes, except that the carbon atoms form extended hollow tubes instead of closed, hollow spheres. Carbon nanotubes can also form as a series of nested, concentric tubes. Carbon nanotubes can be used as nanometer-scale syringe needles for injecting molecules into cells and as nanoscale probes for making fine-scale measurements. Carbon nanotubes can be filled and capped, forming nanoscale test tubes or potential drug delivery devices. Carbon nanotubes can also be "doped," or modified with small amounts of other elements, giving them electrical properties that include fully insulating, semiconducting, and fully conducting.

Carbon Nanotubes: Cellular Automata: an array of identically programmed automata, or "cells," which interact with one another. [David G. Green]

Tiny tubes about 10,000 times thinner than a human hair — consist of rolled up sheets of carbon hexagons. Discovered in 1991 by researchers at NEC, they have the potential for use as minuscule wires or in ultrasmall electronic devices. To build those devices, scientists must be able to manipulate the Nanotubes in a controlled way. Carbon nanotube tips have several advantages [as *atomic force microscopy* tips] , including high aspect ratio for imaging deep and narrow crevices, low tip- sample adhesion for gentle imaging, the ability to elastically buckle rather than break when large forces are applied, and the potential to achieve resolutions in the range of 1.0 nm or less. In addition, carbon nanotubes have well defined molecular structures so that it is possible to control their synthesis to make every tip with an identical structure and resolution. Carbon nanotubes can be selectively modified at their ends with organic or biological molecules to allow functional sensitive *imaging.* As described and developed by Charles Lieber and colleagues, carbon nanotube tips can be 'grown' directly by a process called chemical vapor deposition (CVD), using a reaction of ethylene with an electrodeposited iron catalyst in etched pores on commercial silicon- cantilever- tip assemblies (10). The resulting nanotubes have radii of 3-8 nm if multiwalled; single- walled tubes have smaller radii, on the order of 1-2 nm or less, and potentially less than 0.5 nm if certain conditions are met.

Carbonium ion: A highly reactive cationic chemical species with an even number of electrons and an unoccupied orbital on a carbon atom.

Carbonyl: A chemical moiety consisting of O with a double bond to C. If the C is bonded to N, the resulting structure is termed an amide; if it is bonded to O, it is termed a carboxylic acid or an ester linkage.

Carboxylic acid: A molecule that includes a C having a double bond to O and a single bond to OH.

Catalyst: A chemical species or other structure that facilitates a chemical reaction without itself undergoing a permanent change.

Cathode: The electrode in an electrochemical cell or galvanic couple at which a reduction reaction occurs; in other words the electrode receiving electrons from an external circuit.

Cation: A positively charged ion.

Causality: The law of cause and effect.

cDNA: Abbreviation for complementary deoxyribonucleic acid. cDNA has been copied directly from mRNA, the reverse of transcription. It differs from genomic DNA in the nucleus, since it does not contain the introns, or sequences upstream or downstream of the corresponding genes. cDNA libraries consist of samples from individual tissues, and will only contain the sequences transcribed in that tissue.

Cell engineering: Deliberate artificial modifications to biological cellular systems on a cell-by-cell basis.

Cell pharmacology: Delivery of drugs by medical nanomachines to exact locations in the body.

Cell Repair Machine: A system including nanocomputers and molecular scale sensors and tools, programmed to repair damage to cells and tissues. Molecular and nanoscale machines with sensors, nanocomputers and tools, programmed to detect and repair damage to cells and tissues, which could even report back to and receive instructions from a human doctor if needed.

Cell surgery: In medical nanorobotics, modifying cellular structures using medical nanomachines. Modifying cellular structures using medical nanomachines.

Cell: A membrane-bound unit, typically microns in diameter. All plants and animals are made up of one or more cells (trillions, in the case of human beings). In general, each cell of a multicellular organism contains a nucleus holding all of the genetic information of the organism. (2) A small structural unit, surrounded by a membrane, making up living things.

Ceramic: A nonmetallic material made from clay and hardened by firing at high temperature; it contains minute silicate crystals suspended in a glassy cement.

Cermet: A composite material consisting of a combination of ceramic and metallic meterials. The most common cermets are the cemented carbides, composed of an extremely hard ceramic (e.g. WC, TiC), bonded together by a ductile metal such as cobalt or nickel.

Chemical Vapour Deposition (CVD): a technique used to deposit coatings, where chemicals are first vaporized, and then applied using an inert carrier gas such as nitrogen.

Chemistry: The science (and corresponding technology) that seeks to understand (and control) the atomic and molecular structure of matter.

Chemotactic nanosensor: In medical nanorobotics, a nanosensor used to determine the chemical characteristics of surfaces, possibly configured as a pad coated with an array of reversible, perhaps reconfigurable, artificial molecular receptors.

Chip: A die (unpackaged semiconductor device) cut from a silicon wafer, incorporating semiconductor circuit elements such as a sensor, actuator, resistor, diode, transistor, and/or capacitor. *See* Integrated circuit.

Chromatography: The general name for a series of methods for separating mixtures by employing a system with a mobile phase and a stationary phase.

Chronocyte: In medical nanorobotics, a theorized mobile, mass-storage (nanorobotic) device, similar to a communicyte, that may be used as a mobile source of precisely synchronized universal time inside the human body.

C-kinase: Ca2+-dependent protein kinase that, when activated by diacylglycerol and an increase in the concentration of Ca2+, phosphorylates target proteins on specific serine and threonine residues.

Classical mechanics: Classical mechanics describes a mechanical system as a set of particles (which in a limiting case can form continuous media) having a well-defined geometry at any given time, and undergoing motions determined by applied forces and by the initial positions and velocities of the particles. The forces themselves may have electromagnetic or quantum mechanical origins. Classical statistical mechanics uses the same physical model, but treats the geometry and velocities as uncertain, statistical quantities subject to random thermally-induced fluctuations. Classical mechanics and classical statistical mechanics give a good account of many mechanical properties and behaviors of molecules; but for describing the electronic properties and behaviors of molecules, they are often useless.

CMOS: An acronym for *complementary metal-oxide-semiconductor*, as in CMOS transistor and CMOS logic. Complementary metal oxide semiconductor - integrated circuit containing n-channel and p-channel MOSFETs.

Cobots: Collaborative robots designed to work alongside human operators. Prototype cobots are being used on automobile assembly lines to help guide heavy components like seats and dashboards into cars so they don't damage auto body parts as workers install them. [Wired 5.07 Jargon Watch]

Codons: Organic bases in sets of three that form the genetic code.

Cognotechnology: Convergence of nanotech, biotech and IT, for remote brain sensing and mind control. [Nanodot]

Coherence: From quantum physics, coherence is a property exhibited by matter and energy when it remains unmeasured and is thus able to exist in more than one state simultaneously (superposition).

Col: In describing landforms, a pass between two valleys is sometimes termed a col. In describing molecular potential energy functions, this term is commonly used to describe analogous features of the PES; a col is the region around a saddle point having negative curvature along one axis and positive curvature along all orthogonal axes.

Combinatorial biology: Involves genetic manipulation of bacteria and fungi that produce complex natural products. This technology includes construction of large libraries of recombinant microbes capable of generating novel organic molecules and engineering secondary metabolite biosynthetic pathways to modify valuable biologically active microbial metabolites.

Combinatorial materials design: Uses computing power (sometimes together with massive parallel experimentation) to screen many different materials possibilities to optimize properties for specific applications (e.g., catalysts, drugs, optical materials).

Communicyte: In medical nanorobotics, a theorized mobile, mass-storage (nanorobotic) device that can be used for information transport throughout the human body.

Compliance: The reciprocal of stiffness; in a linear elastic system, displacement equals force times compliance.

Composites: Combinations of metals, ceramics, polymers, and biological materials that allow multi- functional behavior. One common practice is reinforcing polymers or ceramics with ceramic fibers to increase strength while retaining light weight and avoiding the brittleness of the monolithic ceramic. Materials used in the body often combine biological and structural functions (e.g., the encapsulation of drugs).

Composites: Combinations of metals, ceramics, polymers, and biological materials that allow multi- functional behavior. One common practice is reinforcing polymers or ceramics with ceramic fibers to increase strength while retaining light weight and avoiding the brittleness of the monolithic ceramic. Materials used in the body often combine biological and structural functions (e.g., the encapsulation of drugs).

Computational Nanotechnology: Permits the modeling and simulation of complex nanometer-scale structures. The predictive and analytical power of computation is critical to success in nanotechnology: nature required several hundred million years to evolve a functional "wet" nanotechnology; the insight provided by computation should allow us to reduce the development time of a working "dry" nanotechnology to a few decades, and it will have a major impact on the "wet" side as well. [Rice University]

Computronium: A highly (or optimally) efficient matrix for computation, such as dense lattices of nanocomputers or quantum dot cellular automata. [Eugene Leitl]

Conductor: A substance, material or object that allows electricity to flow through it with low resistance. Material such as the metals copper or aluminum that conducts electricity via the motion of electrons.

Configuration space: A mathematical space describing

the three-dimensional configuration of a system of particles (e.g., atoms in a nanomechanical structure) as a single point; the configuration space for an *N* particle system has 3*N* dimensions.

Conformation: The shape of a molecule. A molecular geometry that differs from other geometries chiefly by rotation about single or triple bonds; distinct conformations (termed con formers) are associated with distinct potential wells. Typical biomolecules and products of organic synthesis can interconvert among many conformations; typical diamondoid structures are locked into a single potential well, and thus lack conformational flexibility.

Conjugated: A conjugated pi system is one in which pi bonds alternate with single bonds. The resulting electron distribution gives the intervening single bonds partial double-bond character, the pi electrons become delocalized, and the energy of the system is reduced.

Conjugation: In medical nanorobotics, the docking of two or more nanorobots for the purpose of exchanging information, energy or materials, or to establish a larger multirobotic structure; in biology, the union of two unicellular organisms accompanied by an interchange of nuclear material, as in Paramecium.

Conservative: In design and analysis, a conservative model or a conservative assumption is one that departs from accuracy in such a way that it reduces the chances of a false-positive assessment of the feasibility of the system in question. Conservative assumptions overestimate problems and underestimate capabilities.

Contelligence: (Consciousness + intelligence) The

combination of awareness and computational power required in an Artificially Intelligent network before we could, without loss of anything essential, upload ourselves into them. [Timothy Leary] [AS]

Convergent Assembly: A process of fastening small parts to obtain larger parts, then fastening those to make still larger parts, and so on; convergent assembly can be used to build a product from many, much smaller, components. "...rapidly make products whose size is measured in meters starting from building blocks whose size is measured in nanometers. It is based on the idea that smaller parts can be assembled into larger parts, larger parts can be assembled into still larger parts, and so forth. This process can be systematically repeated in a hierarchical fashion, creating an architecture able to span the size range from the molecular to the macroscopic." [Ralph C. Merkle]

Copolymer: A polymer that consists of two or more dissimilar monomer units in combination in its molecular chains. Also a polymer formed from the polymerization of more than one type of monomer.

Corrosion: Deteriorative loss of a metal as a result of dissolution environmental reactions.

Coulomb [C]: Measure of electrical charge: 1 C is an amount of charge equal to that of about 6.24×10^{18} electrons.

Coupling: A connection between more than one object or energy pathway so that together they function as a single unit.

Covalent bond: A bond formed by sharing a pair of electrons between two atoms. A primary interatomic bond that is formed by the sharing of electrons between neighboring atoms. An attachment between two atoms resulting from the sharing of a pair of electrons

Covalent radius: Given a set of *N* elements that can form covalent single bonds in molecules, with *N*(*N* - 1) possible elemental pairings, it has proved possible to define a covalent radius for each element such that the actual bond length between any two elements that form a covalent single bond is roughly equal to the sums of their covalent radii.

CPU: The central processing unit of a computer, responsible for executing instructions to process information.

Creep: The time-dependent permanent deformation that occurs under stress; for most materials it is important only at elevated temperatures.

Crosslinked polymer: A polymer in which adjacent linear molecular chains are joined at various positions by covalent bonds.

Cross-linking: A process forming chemical bonds between two separate molecular chains.

Cross-sensitivity: The influence of one measurand on the sensitivity of a sensor, another measurand.

Crosstalk: Electromagnetic noise transmitted between leads or circuits in close proximity to each other.

Crybiology: The science of biology at low temperatures; research in cryobiology has made possible the freezing and storing of sperm and blood for later use.

Crystal Lattice: The regular three-dimensional pattern of atoms in a crystal.

Crystal structure: For crystalline materials, the manner in which atoms or ions are arrayed in space. It is defined in terms of the unit cell geometry and the atom positions within the unit cell.

Crystallescence: In medical nanorobotics, the crystallization of solid solute that is offloaded by nanorobot sorting rotors at a concentration that exceeds the solvation capacity of the surrounding solvent.

Curie temperature (also Curie point) (T_c): The temperature above which a ferromagnetic or ferrimagnetic material becomes paramagnetic. For iron the Curie point is 760oC and for nickel 356°C.

Current [A]: Measure of rate of flow of electric charge: a one-ampere current is a flow of 1 C of charge per second.

Cutoff: Condition in a diode or bipolar junction transistor in which the potential across a p-n junction prevents current flow.

Cyclic: A structure is termed cyclic if its covalent bonds form one or more rings.

Cycloaddition: A reaction in which two unsaturated molecules (or moieties within a molecule) join, forming a ring.

Cyclotron: A type of particle accelerator in which an ion introduced at the center is accelerated in an expanding spiral path by use of alternating electrical fields in the presence of a magnetic field.

Cytocarriage: In medical nanorobotics, the commandeering of a natural motile cell, by a medical nanorobot, for the purposes of in vivo transport (of the nanorobot), or to perform a herding function (of the affected cell), or for other purposes.

Cytocide: The killing of living cells.

Cytography: A physical description (and mapping) of the living cell.

Cytoidentification: Identification of cell type.

Cytometrics: The quantitative measurement of cell sizes, shapes, structures, and numbers.

Cytopenetration: In medical nanorobotics, entry into cells by penetrating the plasma membrane.

Cytoskeletolysis: In medical nanorobotics, purposeful destruction of the cellular cytoskeleton by a nanorobot, for cytocidal purposes. (2) Tomographic imaging of an individual cell.

Cytovehicle: In medical nanorobotics, a living cell that has been commandeered by a medical nanorobot for use during cytocarriage.

D

Dalton: Unit of molecular mass. Approximately equal to the mass of a hydrogen atom (1.66 x 1E-24 g).

ĐBERGOO: A related term to grey goo, used (jokingly) to refer to the mistaken idea that during the singularity powerful technologies would decimate non-transhumanists, and that some transhumanists would see this as desirable (which is clearly against theTranshuman Principles). [Dale Carrico 1996]

Debye Shielding: The Debye length in front of a sensing electrode depends on the ionic strength of the electrolyte used. In a 0.001N NaCl the Debye length measures 96.5 Å, while for a 1.0 N solution it is reduced to 3.0 Å. An adsorbed protein can stick out from the surface for as much as 50 to 100 Å. As a result, the charges which could contribute to the surface potential will be shielded in a 1.0 N solution. To make more sensitive measurements a solution of low ionic strength should be used.

Degenerate: Not a moral judgment but an adjective that describes multiple states that amount to the same thing: different triplet combinations of nucleotide bases (codons) that code for the same amino acid, for example.

Degradation: A term used to describe the deteriorative processes that occur with polymeric materials, including swelling, dissolution, and chain scission. In medical nanorobotics, a crude form of functional navigation in which artificial conditions detectable by in vivo medical nanorobots are created at or near the target treatment site, such as warm or cold spots, pressure spots, or injected chemical plumes. Dramatic change in conformation of a protein or nucleic acid caused by heating or by exposure to chemicals and usually resulting in the loss of biological function.

Denaturation: The breaking down of the three-dimensional structure of a protein resulting in the loss of its function.

Dendrimer: A dendrimer is a spherical, highly branched polymer molecule. Dendrimers are made from two different monomers – a reactive amine and an acrylic acid – and are assembled in discrete steps, which allow them to be constructed with an exact size that depends on the number of enlargement steps. This picture shows a dendrimer that has undergone three "generations" of enlargements.

Dendrimers are of particular interest for cancer applications because of their defined and reproducible size, but more importantly, because it is easy to add a variety of other molecules to the surface of a dendrimer. Such molecules could include tumor-targeting agents (including but not restricted to monoclonal antibodies), imaging contrast agents to pinpoint tumors, drug molecules for delivery to a tumor, and reporter molecules that might detect if an anticancer drug is working. (2) A dendrimer is a tree-like highly branched polymer molecule (Greek *dendra = tree*). Dendrimers are synthesized from monomers with new branches added in discrete steps ("generation") to form a tree-like architecture. A high level of synthetic control is achieved through step-wise reactions and purifications at each step to control the size, architecture, functionality and monodispersity. Several different kinds of dendrimers have been synthesized utilizing different monomers and some are commercially available. This picture shows a "3rd generation" polyamidoamine (PAMAM) dendrimer.

Dendrimers are of particular interest for cancer applications because of their defined and reproducible size, but more importantly, because it is easy to attach a variety of other molecules to the surface of a dendrimer. Such molecules could include tumor-targeting agents (including but not restricted to monoclonal antibodies), imaging contrast agents to pinpoint tumors, drug molecules for delivery to a tumor, and reporter molecules that might detect if an anticancer drug is working.

Dendrimer: A dendrimer is a tree-like highly branched polymer molecule (Greek *dendra = tree*). Dendrimers are synthesized from monomers with new branches added in discrete steps ("generation") to form a tree-like architecture. A high level of synthetic control is achieved through step-wise reactions and purifications at each step to control the size, architecture, functionality and monodispersity. Several different kinds of dendrimers have been synthesized utilizing different monomers and some are commercially available. This picture shows a "3rd generation" polyamidoamine (PAMAM) dendrimer.

Dendrimers are of particular interest for cancer applications because of their defined and reproducible size, but more importantly, because it is easy to attach a variety of other molecules to the surface of a dendrimer. Such molecules could include tumor-targeting agents (including but not restricted to monoclonal antibodies), imaging contrast agents to pinpoint tumors, drug molecules for delivery to a tumor, and reporter molecules that might detect if an anticancer drug is working. (2) A polymer having a regular branched structure; If suitably functionalized may be used as a soluble support, in which case the desired, dendrimer- supported, material may be isolated by size- exclusion chromatography. Dendrimers may also be attached to a polymer and used as a **solid support**, with significantly increased loading over the initial resin. [IUPAC COMBINATORIAL CHEMISTRY]

Dendrimers consist of interconnected monomeric subunits that hybridize to form a tree- like structure. Each monomer is a double- stranded DNA molecule where the two strands share a region of sequence complementarity in the middle of molecule. [CHI Microarrays report]

A dendrimer (from Greek *dendra* for tree) is an artificially manufactured or synthesized molecule built up from branched units called monomers. Such processes involve working on the scale of nanometers (a nanometer is 10^{-9} meter or a millionth of a millimeter). Technically, a dendrimer is a **polymer**, which is a large molecule comprised of many smaller ones linked together. Dendrimers have some proven applications, and numerous potential applications. [whatis.com] Also known as *"cascade molecules"* (3) From the Greek word *dendra* - tree, a dendrimer is polymer that branches.

Dendrimers: From the Greek word dendra - tree, a dendrimer is polymer that branches. "...a tiny molecular structure that interacts with cells, enabling scientists to probe, diagnose, cure or manipulate them on a nanoscale." Invented by Professor Donald Tomalia from Central Michigan University. [SmallTimes] See this article for a great explanation Dendrimers: Branching out into new realms of molecular architecture.

Dendrite: Extension of a nerve cell, typically branched and relatively short, that receives stimuli from other nerve cells

Deoxyribonucleic acid (DNA): A huge nucleotide polymer having a double helical structure with complementary bases on two strands. Its major functions are protein synthesis and the storage and transport of genetic information.

Deoxyribonucleic acid: Deoxyribonucleic acid are long chains consisting of four kinds of nucleotides; the order of these nucleotides encodes the information needed to construct protein molecules. These in turn make up much of the molecular machinery of the cell. DNA is the genetic material of cells.

Dermal Zippers: In medical nanorobotics, a theorized medical nanorobot that can rapidly perform incision-wound repairs to the dermis and epidermis; dermal zippers.

Design Ahead: The use of known principles of science and engineering to design systems that can only be built with tools not yet available; this permits faster exploitation of the abilities of new tools.

Design Diversity: A form of redundancy in which components of different design serve the same purpose; this can enable systems to function properly despite design flaws.

Design: To plan and delineate with an end in mind and subject to constraints.

Desmosome: Specialized cell-cell junction, usually formed between two epithelial cells, characterized by dense plaques of protein into which intermediate filaments in the two adjoining cells insert.

Detergent: Type of small amphipathic molecule that tends to coalesce in water, with its hydrophobic tails buried and its hydrophilic heads exposed; widely used to solubilize membrane proteins.

Determination: Commitment by an embryonic cell to a particular specialized path of development; it reflects a change in the internal character of the cell.

Development: Succession of changes that take place in an organism as a fertilized egg gives rise to an adult plant or animal.

Devitrification: The process in which a glass (noncrystalline or vitreous solid) transforms to a crystalline solid.

Dialysis: A phenomenon in which a semipermeable membrane allows transfer of both solvent molecules and small solute molecules and ions.

Diamagnetism: A weak form of induced or nonpermanent magnetism for which the magnetic susceptibility is negative. A type of magnetism associated with paired electrons, that causes a substance to be repelled from the inducing magnetic field.

Diamondoid: Structures that resemble diamond in a broad sense, strong stiff structures containing dense, three dimensional networks of covalent bonds; diamondoid materials could be as much as 100 to 250 times as strong as titanium, and far lighter.

Diamondoid: Stuctures that resemble diamond in a broad sense, strong stiff structures containing dense, three dimensional networks of covalent bonds, formed chiefly from first and second row atoms with a valence of three or more. Many of the most useful diamondoid structures will in fact be rich in tetrahedrally coordinated carbon. [NTN] Materials with superior strength to weight ratio, as much as 100 to 250 times as strong as Titanium, and much lighter. Possibly used to build stronger lighter rockets and space components, or a variety of other earth-bound articles for which weight and strength are a consideration.

Diamondophagy: Eating diamond.

Die: A die (unpackaged semiconductor device) cut from a silicon wafer, incorporating semiconductor circuit elements such as a sensor, actuator, resistor, diode, transistor, and/or capacitor. *See* Integrated circuit.

Dielectric (breakdown) strength: The magnitude of an electric field necessary to cause significant current passage through a dielectric material.

Dielectric constant (å): The ratio of the permittivity of a medium to that of a vacuum. Also called the relative dielectric constant or relative permittivity.

Dielectric displacement: The magnitude of charge per unit area of capacitor plate.

Dielectric: Any material that is electrically insulating.

Diffusion coefficient: The constant of proportionality between the diffusion flux and the concentration gradient in Fick's first law. Its magnitude is indicative of the rate of atomic diffusion.

Diffusion: A thermochemical process whereby controlled dopants are introduced into a substrate.

Digital: Refers to systems employing only quantized (discrete) states to convey information (also see"analog").

Dimer: A molecule formed by the joining of two identical monomers.

Diode: two-terminal device that conducts current well in one direction and poorly in the other.

Dip Pen Nanolithography (DPN): A method for nanoscale patterning of surfaces by the transfer of a material from the tip of a scanning probe microscope onto the surface. (2) New AFM- based soft- lithography technique which was invented in our labs. Dip-Pen Nanolithography (DPN) is a scanning probe nanopatterning technique in which an AFM tip is used to deliver molecules to a surface via a solvent meniscus, which naturally forms in the ambient atmosphere. (3) An AFM-based soft-lithography technique. See example at Surface science in the Mirkin Group

Dip: Dual In-line Package - common ceramic or plastic enclosure for an integrated circuit.

Dipole (electric): A pair of equal yet opposite electrical charges that are separated by a small distance.

Directed-Assembler: A specific type of assembler that makes use of directed-assembly, such that the assembly process requires external energy or information input.

Disassembler: A system of nanomachines able to take an object apart a few atoms at a time, while recording its structure at the molecular level. (2) An instrument able to take apart structures a few atoms at a time, recording structural information at each step. (3) In molecular nanotechnology, a nanomachine or system of nanomachines able to take an object apart while at each step recording the structure and composition of that object at the molecular level.

Disasterbation: Idly fantasizing about possible catastrophes (ecological collapse, full-blown totalitarianism) without considering their likelihood or considering their possible solutions and preventions. [David Krieger, 1993]

Disequilibration: In medical nanorobotics, maintenance or inducement of a state of perpetual ionic, chemical, or energetic disequilibrium in a living cell by a medical nanorobot, usually for the purpose of inducing cytocide.

Dislocation: A linear crystalline defect around which there is atomic misalignment. Plastic deformation corresponds to the motion of dislocations in response to an applied shear stress. Edge, screw, and mixed dislocations are possible.

Disruptive Technology: Technology that is significantly cheaper than current, is much higher performing, has greater functionality, and is frequently more convenient to use. Will revolutionize markets by superseding existing technology. "Paradigm shifting" is a well-worn connotation. Although the term may sound negative to some, it is in fact neutral. It is only negative when businesses who are unprepared for change fail to adapt, only to fall behind and fail. The results are not *evolutionary*, they are *revolutionary*.

Disruptive techonlogies: In his book, *The Innovator's Dilemma*, Harvard professor Clayton Christensen used the term "disruptive technologies" to describe developments capable of displacing established market leaders.

Dissolution: Deterioration in an organism such that its original structure cannot be determined from its current state.

Distributed Intelligence: An intelligent entity which is distributed over a large volume (or inside another system, like a computer network) with no distinct center. This is the opposite to the strategy of Concentrated intelligences. Distributed intelligences have much longer communications lags, but are more flexible in their structure and can survive damage to their parts.

DNA (Deoxyriobonucleic Acid): DNA molecules are long chains consisting of four kinds of nucleotides; the order of these nucleotides encodes the information needed to construct protein molecules. These in turn make up much of the molecular machinery of the cell. DNA is the genetic material of cells. (See also RNA.)

DNA Chip: Also: Gene Chip and DNA Microchip. A purpose built microchip used to identify mutations or alterations in a gene's DNA. See DNA Chip Technology

DNA diagnostics - miniaturization of: In the areas of sample preparation and assay, it is clear that miniaturization is key. To reduce the size of samples by a factor of 10 or greater, barriers in microfluidics, micromachining, robotics, microchemistry, nucleic acid chemistry, and surface chemistry must be overcome. To implement miniaturized protocols accurately and efficiently, substantial automation of the process will be required. In the development of miniaturized systems, it is essential that the system can be adapted for high levels of parallelization.

Miniaturization poses significant technological risks. Currently, there exists no universally accepted precedent for the handling, replication, amplification, or cloning of DNA in nanoliter volumes. Due to the size and charge of the DNA molecule, and the relative instability of many of the enzymes involved in the sample preparation processes, nanoliter and less volumes may pose substantial challenges. In addition, interactions of the biological molecules with the surfaces of the reaction chambers must be minimized. For some methodologies, it is not clear what the optimal sample will be, so substantial improvement in DNA fragmentation technologies or DNA cloning vectors may be required for the ultimate efficient application to diagnostics.

Improvements in any of these areas are likely to be of value to other non- DNA based diagnostic applications such as antibody screening protocols and enzyme based diagnostics, because miniaturized robotic or micro- electro mechanical systems developed for DNA could be modified to be used for these purposes.

DNA nanotechnology: Ned Seeman, DNA Nanotechnology, New York Univ., US, 2003

DNA Probes: A DNA or nucleic acid probe is a short strand of DNA that locates and binds to its complementary sequence in samples containing single strands of DNA or RNA enabling identification of specific sequences.

Nucleic acid probe assays exploit the fundamental hybridization reaction that occurs spontaneously between two complementary DNA:DNA or DNA:RNA strands. As in immunoassays, detection of the hybrid requires that the probe be labeled. Various direct and indirect methods have been devised for the detection of the hybrid. Direct labeling involves attaching the label directly to the probe sequence; indirect labeling binds an antibody to the DNA:DNA or DNA:RNA hybrid.

As in immunoassays, non-isotopically-labeled probes are preferred over radio-labeled probes primarily because of radiation hazards, disposal problems, and short reagent shelf life. In addition, the factors determining the detection limits of hybridization assays based on labeled probes are similar to those in immunoassays. Therefore, the development of a simple, inexpensive and sensitive direct detection system which eliminates the use of labels is highly desirable.

DNA Sequencing: There are two main classical methods for sequencing DNA: The first method, developed by Allan Maxam and Walter Gilbert, involves chemicals used to cleave the DNA at certain positions, generating a set of fragments that differ by one nucleotide.

The second method, developed by Fred Sanger and Alan Coulson, involves enzymatic synthesis of DNA strands that terminate in a modified nucleotide.

Analysis of fragments is similar for both methods and involves gel electrophoresis and autoradiography or fluorescence. The enzymatic method has largely replaced the chemical method as the technique of choice, although there are some situations where chemical sequencing can provide data more easily than the enzymatic method.

DNA: A molecule encoding genetic information, found in the cell's nucleus. Deoxyribose Nucleic Acid. DNA is a molecule used in biological systems to store genetic information. It has a long, double stranded structure with the two strands wrapped around each other to form a right handed helix. There are about 10 nucleotide pairs per helical turn. Each strand is composed of a sugar phosphate backbone and attached bases.

It is connected to a complementary strand by hydrogen bonding (non- covalent) between paired bases, adenine (A) with thymine (T) and guanine (G) with cytosine (C). DNA is an abbreviation of DeoxyriboNucleic Acid. : It is the genetic 'blueprint' of all life on this planet. : DNA typing is often applied to Human Identification which has also been called DNA Fingerprinting. DNA Fingerprinting has been attributed to Alex Jeffries, who with his colleagues in 1985, discovered a sequence of nucleotides, subsequently termed short tandem repeats (STR) when comparing the DNA sequence of the human á-globin chromosomes isolated from different individuals.

Dnhancer: Regulatory DNA sequence to which gene regulatory proteins bind, influencing the rate of transcription of a structural gene that can be many thousands of base pairs away.

Domain: A region of a ferromagnetic or ferrimagnetic material in which all atomic or ionic magnetic moments are aligned in the same direction.

Dopeyballs: Superconducting Buckyballs (they) have the highest critical temperature of any known organic compound. see The Buckyball Collection [Florida St U]

Doping: Process of introducing impurity atoms into a semiconductor to affect its conductivity.

DRAM: Dynamic Random Access Memory — memory in which each stored bit must be refreshed periodically.

Drift: Gradual departure of the instrument output from the calibrated value. An undesired slow change of the output signal.

Dry Nanotechnology: Derives from surface science and physical chemistry, focuses on fabrication of structures in carbon (e.g. fullerenes and nanotubes), silicon, and other inorganic materials. Unlike the "wet" technology, "dry" techniques admit use of metals and semiconductors.

The active conduction electrons of these materials make them too reactive to operate in a "wet" environment, but these same electrons provide the physical properties that make "dry" nanostructures promising as electronic, magnetic, and optical devices. Another objective is to develop "dry" structures that possess some of the same attributes of the self-assembly that the wet ones exhibit. [Rice University]

DSP: Digital Signal Processing; a process by which a sampled and digitized data stream (real-time data such as sound or images) is modified in order to extract relevant information. Also, a digital signal processor.

Ductility: A measure of a material's ability to undergo plastic deformation before fracture; expressed as percent elongation (%EL) or percent area reduction (%AR) from a tensile test.

DumbSizing: Apealing to the least common denominator by explaining difficult concepts in such a manner so they loose meaning. Also, talking down to someone less informed or learned. [uhf]

Dynamic characteristics: A description of an instrument's behavior between the time a measured quantity changes value and the time the instrument obtains a steady response.

Dynamic error: The error that occurs when the output does not precisely follow the transient response of the measured quantity.

Dynamic range: The ratio of the largest to the smallest values of a range, often expressed in decibels.

Dyson Scenario, the: Life expands into the universe, which is open. As the universe cools, life stores energy to survive (do information processing). It waits until the universe is cool enough, performs some processing with part of its energy stores, then waits until the universe has cooled so much that the remaining energy can be used to do an equal amount of computation, and so on.

Essentially life has to adapt as the universe grows older, changing itself to be able to survive when the stars grow cold. If the universe is open, there will be plenty of time to work in, but energy will become very scarce. Dyson has shown that a finite amount of energy is enough to guarantee infinite survival if it is spent sufficiently slowly.

Dyson Sphere: A shell built around a star to collect as much energy as possible, originally proposed by Freeman Dyson (although he admits to have borrowed the concept from Olaf Stapledon's novel *Star Maker* (1937)). In the original proposal the shell consists of many independent solar collectors and habitats in separate orbits (also known as a Type I Dyson Sphere), but later people have discussed rigid shells consisting of only one piece (called a Type II Dyson Sphere).

The latter construction is unfortunately both unstable (since it will experience no net attraction of the star), requires super-strong materials and have no internal gravity. The Dyson Sphere is a classic example of mega-technology and common in Science Fiction. See also The Dyson Sphere FAQ.

Dystopia: Often used to describe a society where people lead dehumanized, fearful, and technology-restricted lives. In other words, a totalitarianism or theocracy, where books are burned, reading of dangerous ideas is proscribed, and the state controls science.

E

Ecophagy: (or Global Ecophagy) Consuming the biological environment. Coined and defined by Robert A. Freitas Jr. (Research Scientist Zyvex Corp). Frequently associated with "gray goo," as ecophagy (uncontrolled self-replication) is its main prupose. See "Some Limits to Global Ecophagy by Biovorous Nanoreplicators, with Public Policy Recommendations" where Dr. Freitas said "Perhaps the earliest-recognized and best-known danger of molecular nanotechnology is the risk that self-replicating nanorobots capable of functioning autonomously in the natural environment could quickly convert that natural environment (e.g., "biomass") into replicas of themselves (e.g., "nanomass") on a global basis, a scenario usually referred to as the 'gray goo problem' but perhaps more properly termed 'global ecophagy.'"

Ecosystem protector: A nanomachine for mechanically removing selected imported species from an ecosystem to protect native species.

EDP: Ethylene diamine pyrocatechol.

Effervescence: In medical nanorobotics, bubble formation by a gaseous solute that is offloaded by nanorobot sorting rotors at a concentration that exceeds the solvation capacity of the surrounding solvent.

EI - Emergent Intelligence: An intelligent system that gradually emerges from simpler systems, instead of being designed top down.

Elastic deformation: A nonpermanent deformation that totally recovers upon release of an applied stress.

Elastomer: A polymeric material that may experience large and reversible elastic deformations.

Electric field [V/m]: In simplest form, the potential difference between two points divided by the distance between the two.

Electrical Bistability: A phenomenon in which an object exhibits two states of different conductivity at the same applied voltage.

Electrical breakdown: Condition in which, particularly with high electric field, a nominal insulator becomes electrically conducting.

Electrochemical gradient: Driving force that causes an ion to move across a membrane due to the combined influence of a difference in its concentration on the two sides of the membrane and the electrical charge difference across the membrane.

Electroluminescence: In electrical engineering: the emission of visible light by a p-n junction across which a forward-biased voltage is applied. In electrochemistry: emission of light by a molecule which is being reduced or oxidized on a biased electrode. If the exciting cause is a photon, rather than an electron, the process is called photoluminescence.

Electrolyte/Insulator/Silicon (EIS): Structures at the heart of a broad family of potentiometric silicon sensors. The best-known member of the family is the ion-sensitive field effect transistor, known as the ISFET or CHEMFET and the light-addressable potentiometric sensor LAPS . The principle of operation of devices using such structures is as follows. A potential with respect to a reference electrode is generated at the interface between the liquid solution and the insulator. The surface potential ($ø_0$) is determined by that ionic species which has the fastest exchange rate (i_o) with the membrane covering the insulator. If no intentional membrane is deposited on an oxide covered insulator that species will be H^+.

Surface potential changes in turn change the Si flat-band voltage V_{FB}. The flat-band voltage is the potential one needs to apply to the Si in order to have the bands flat throughout the semiconductor. The flat band voltage of an EIS structure has been shown to be given by: $V_{FB} = E_{REF} - Ö^{Si}/q - ø_0 - Q_{ins}/C_{inss}$ where V_{FB} stands for the flat-band voltage of the structure, E_{REF} for the reference electrode potential, $Ö^{Si}$ for the work function of silicon, $ø_0$ for the surface potential at the insulator/electrolyte interface, Q_{ins} for the charge at the insulator/silicon interface and C_{ins} for the insulator capacitance. At least two terms in the above equation are not known with a precision greater than a few hundred millivolts. This is true for E_{REF} as well as for Q_{ins}/C_{ins} which can vary from device to device by several hundred millivolts. For a given EIS sensor, these inaccurately known quantities are constant, and variations in flat-band voltage can be equated to variations of the surface potential.

Electrolyte: A solution through which an electric current may be carried by the motion of ions.

Electromotive force (emf) series: A series of chemical elements arranged in order of their electromotive force. The electromotive force is the greatest potential difference that can be generated by a particular source of electric current. In practice this potential may be observable only when the source is not supplying current, because of its internal resistance.

Electron acceptor: Atom or molecule that takes up electrons readily, thereby gaining an electron and becoming reduced.

Electron Beam Lithography (EBL): A method of fabricating sub-micron and nanoscale features by exposing electrically sensitive surfaces to an electron beam.

Electron beam: A stream of electrons moving in the same direction at the same speed.

Electron carrier: Molecule such as cytochrome c that transfers an electron from a donor molecule to an acceptor molecule.

Electron donor: Molecule that easily gives up an electron, becoming oxidized in the process.

Electron: Electrons are negatively charged elementary particles with a mass of $9.1093897 \times 10^{-31}$ kg, a charge of $-1.60217733 \times 10^{-19}$ C, and spin 1/2. They cannot be divided into smaller particles. What is the radius of an electron? There is not an easy answer to this question. In an atom, the wavefunction of an electron has a width of about 10^{-10} meters. The wavefunction for an electron in a metal or travelling through vacuum can be much larger; it can be meters in width. Elementary negative particle whose charge is 1.602 x10^{-19} coulombs.

Electron state (level): One of a set of discrete, quantized energies that are allowed for electrons. In the atomic case each state is specified by 4 quantum numbers.

Electron Transport Chain: Biomolecular machinery present in prokaryotic membranes and eukaryotic mitochnodria that couples the flow of electrons to proton pumps in order to convert energy from sugar to ATP.

Electron transport: Movement of electrons from a higher to a lower enery level along a series of electron carrier molecules, as in oxidative phosphorylation and photosynthesis.

Electron-beam lithography: Electron beam lithography is a process that can be used to write fine patterns with an electron beam. An electron beam can be generated by extracting electrons from a sharp needle with an electric field. If the electrons are then accelerated towards a metal plate with a small hole in it, a narrow beam of electrons emerges from this hole.

The electrons are typically accelerated through a potential of several kilovolts and are traveling at a good fraction of the speed of light. An electron beam can be deflected by electric and magnetic field which makes it possible to write with the beam. Such electron beams are used in televisions and computer monitors. It is possible to focus an electron beam to have a very small diameter of about 1 nanometer. Such narrowly focused beams are used in scanning electron microscopes, transmission electron microscopes, and electron-beam pattern generators. Typically in an electron-beam pattern generator, the electron beam writes a pattern in a thin organic film that is coated over a wafer. Usually the wafer is a single crystal of silicon that is about 0.5 mm thick and 10 - 30 cm in diameter. The thin (~100 nm) organic film contains long molecules that wrap around each other like cooked spagetti.

The energetic electrons in the electron beam cut these molecules up into small pieces. The film is then dipped in developer which disolves the short pieces but leaves the long unexposed sections unaffected. A pattern is thereby defined in the resist. The pattern can be transfered to the substrate by putting the wafer in a gas or liquid that dissolves the substrate material. Patterns with features of a few tens of nanometers can be made this way.

Electronegative: Describing elements that tend to gain electrons and form negative ions. The halogens are typical electronegative elements.

Electronvolt: An electronvolt is defined as 1.6022×10^{-19} Joules. The energy needed to move an electron from a metal electrode at voltage V_1 to an electrode at voltage V_2 is $V_1 - V_2$ electron volts. One electronvolt is about the amount of energy needed to dissociate a molecule.

Elongation factor: Protein required for the addition of amino acids to growing polypeptide chains on ribosomes.

Embryogenesis: Development of an embryo from a fertilized egg, or zygote.

Emergence: A complex whole created by simple parts, as in the brain where billions of neurons work individually, but collectively make up our consciousness and give us the ability to think, rationalize, and create.

EMI: Electromagnetic interference.

Emulation: An absolutely precise simulation of something, so exact that it is equivalent to the original (for example, many computers emulate obsolete computers to run their programs). [AS] The Star Trek replicator is an example.

Enabling science and technologies: Areas of research relevant to a particular goal, such as nanotechnology. Also, technology that "enables" other technology to advance, such as the transistor enabled the computer chip revolution, as did photolithography.

Endocrine cell: Specialized animal cell that secretes a hormone into the blood; usually part of a gland, such as the thyroid or pituitary gland.

Endocytosis: Uptake of material into a cell by an invagination of the plasma membrane and its internalization in a membrane-bounded vesicle. (See also pinocytosis and phagocytosis.)

Endoplasmic reticulum (ER): Labyrinthine, membrane-bounded compartment in the cytoplasm of eucaryotic cells, where lipids are synthesized and membrane-bound proteins are made.

Endosome: Membrane-bounded organelle in animal cells that carries materials newly ingested by endocytosis and passes many of them on to lysosomes for degradation.

Endothelium: Single sheet of highly flattened cells (endothelial cells) that forms the lining of all blood vessels. Regulates exchanges between the bloodstream and surrounding tissues and is usually surrounded by a basal lamina.

Energy [J]: Capacity for performing work or to cause heat flow. Like work itself, it is measured in Joules.

Energy Conversion: Changing energy from one form to another.

Engineering: The use of scientific knowledge and trial-and-error to design systems. (See Science.)

Engulf formation: In medical nanorobotics, a configuration that may be adopted by a metamorphic nanorobot, in which the nanorobot reshapes itself to create an interior cavity capable of trapping a living cell, virion, or other biological particle.

Entanglement: From quantum mechanics, entanglement is a relationship between two systems in which they exist in more than one state simultaneously (by superposition), and the state of one system determines the state of the other. (2) From quantum mechanics, entanglement is a relationship between two objects in which they both exhibit superposition but once the state of one object is measured, the state of the other is also known.

Enthalpy (H): A property of a system equal to E + PV, where E is the internal energy of the system, P is the pressure of the system, and V is the volume of the system. At constant pressure the change in enthalpy equals the energy flow as heat.

Entropy: A measure of the disorder of a closed system. The second law of thermodynamics states that the entropy (and disorder) increases as time moves forward. Thermodynamic quantity that measures the degree of disorder in a system; the higher the entropy, the more the disorder.

Enzyme immunoassay/EIA: In an enzyme immunoassay (EIA), an enzyme-labeled antibody or antigen is used for the detection and quantification of the antigen-antibody reaction. In an electrochemical EIA, the enzyme-catalyzed reaction is monitored electrochemically (amperometric, potentiometric, voltametric or conductometric). In EIA, the antibody-antigen reaction furnishes the needed specificity. The enzyme label provides the sensitivity via chemical amplification. A large molecule, usually a protein, that catalyzes biological reactions. A protein that acts as a catalyst in a biochemical reaction. Protein that catalyzes a specific chemical reaction.

Enzymes: Molecular machines found in nature, made of protein, which can catalyze (speed up) chemical reactions.

Epitaxial or epi: A single-crystal semiconductor layer grown upon a single-crystal substrate having the same crystallographic characteristics as the substrate material.

EPROM: Electrically Programmable Read-Only Memory — nonvolatile memory device.

Eurisko: A computer program developed by Professor Douglas Lenat which is able to apply heuristic rules for performing various tasks, including the invention of new heuristic rules.

EviDots-MP Functionalized for Life Sciences: Evident then further envelopes the plated InGaP quantum dots with a natural lipid coating. This coating allows for high stability in water, can be readily functionalized for life science applications, is long lasting in the harsh environments encountered in biological assays, and is less toxic to cells than traditional heavy metal quantum dots.

Evolution: A process in which a population of self-replicating entities undergoes variation, with successful variants spreading and becoming the basis for further variation.

Exploratory engineering: Design and analysis of systems that are theoretically possible but cannot be built yet, owing to limitations in available tools.

Exponential assembly: A manufacturing architecture starting with a single tiny robotic arm on a surface. This first robotic arm makes a second robotic arm on a facing surface by picking up miniature parts ó carefully laid out in advance in exactly the right locations so the tiny robotic arm can find them ó and assembling them. The two robotic arms then make two more robotic arms, one on each of the two facing surfaces. These four robotic arms, two on each surface, then make four more robotic arms. This process continues with the number of robotic arms steadily increasing in the pattern 1, 2, 4, 8, 16, 32, 64, etc. until some manufacturing limit is reached (both surfaces are completely covered with tiny robotic arms, for example). This is an exponential growth rate, hence the name exponential assembly. [ZY] See Exponential Assembly

Exponential general-purpose molecular manufacturing: The word exponential refers to the rapid pace — probably unprecedented — at which this technology may be deployed. A compact automated molecular manufacturing system will be able to make more manufacturing systems. We're talking about factories that can build duplicate factories — and probably do it in less than a single day. The math is simple: if one factory makes two, and two factories make four, then within ten days you could have one thousand factories, in ten more days a million factories, and ten days after that a billion factories. Within the span of just a few weeks, in theory, every household in the world could have one of their own, to make most of the products they need, at just the cost of raw materials.

Exponential general-purpose molecular manufacturing means a manufacturing system capable of making a wide range of technologically advanced products, far superior to what we have today, much cheaper, much faster, and able to multiply its own source of production exponentially.

Exponential Growth: Growth that proceeds in a manner characterized by periodic doublings. Inaccurately referred to as "self-replication," exponential growth refers to the process of growth or replication involving doubling within a given period.

Extrinsic: Characterizes doped, rather than pure, semiconductor.

F

Fab: For "fabrication", a term referring to the making of semiconductor devices such as microprocessors.

Fabricator: A small nano-robotic device that can use supplied chemicals to manufacture nanoscale products under external control. Fabricators could work together to build macroscale products by convergent assembly. Similar to assemblers, but less complex, easier to build, and probably more efficient.

Face-centered cubic: A crystal structure found in common elemental metals also FCC. Within the cubic unit cell, atoms are located at all corner and face-centered positions.

Fact Forum: A procedure for seeking facts through a structured, arbitrated debate between experts.

Fail-stop: Describes a component or subsystem that, in the event of a failure, produces no output (e.g., of material or data) rather than producing a damaged or incorrect output.

Farad: The unit of capacitance (see"capacitance").

Faraday: A constant representing the charge on one mole of electrons; 96, 485 coulombs.

Far-Field Diffraction Limit: A physical limit to how focused a beam of light can be.

Fatty acid: Compound such as palmitic acid that has a carboxylic acid attached to a long hydrocarbon chain. Used as a major source of energy during metabolism and as a starting point for the synthesis of phospholipids.

Fault-tolerant: Describes a system that can suffer failure in a component or subsystem, yet continue to function correctly.

FEA: Finite element analysis.

Femtoengineering: Will involve engineering using mechanisms within a quark.

Femtometer: [abbr: fm] A unit suitable to express the size of atomic nuclei. One quadrillionth (10 to minus 15) of a meter.

Femtosecond: Femtosecond is one quadrillionth of a second, and is to a second what a second is to 32,700,000 years. At 186,000 miles per second, in one femtosecond light travels only far enough to traverse about 1,000 silicon atoms. When used to time a laser pulse, it allows for ultra-precise micromachining, with virtually no damage to surrounding material.

Femtotechnology: Femtotechnology the art of manipulating materials on the scale of elementary particles (leptons, hadrons, and quarks). [CA-B] The next step smaller after picotechnology, which is the next step smaller after nanotechnology.

Fermi energy: The energy level in a solid at which the probability of finding an electron is 1/2. For a metal, the energy corresponding to the highest filled electron state in the valence band at 0 K.

Fermi Paradox: "If there are other intelligent beings in the Universe, why aren't they here?" Since it appears to be quite possible for a technological species to spread across the galaxy in less than 10 million years (using von Neumann machines) or otherwise change things on such a large scale that it would be very visible (see Kardaschev types), the lack of such evidence is puzzling or implies that other technological civilizations doen't exist. There have been many attempts to explain this, for example the "Wildlife Preserve" idea (the aliens doesn't want to interfere with younger civilizations), that they transcend and become incomprehensible, that they hide or that they are actually here, hidden on the nanoscale, but the problem with these attempts is that most of them just explain why some aliens would not be apparent. [E. Fermi]

Ferroelectric material: A dielectric material such as Rochelle salt and barium titanate with a domain structure containing dipoles (asymmetric distributions of electrical charge) which spontaneously align. Their domain structure makes them analogous to ferromagnetic materials. They exhibit hysteresis and usually the piezoelectric effect.

Ferromagnetism: Permanent and large magnetizations found in some metals (e.g. Fe, Ni and Co), resulting from the parallel alignment of neighboring magnetic moments.

Fertilization: Fusion of a male and a female gamete (both haploid) to form a diploid zygote, which develops into a new individual.

FET: Field-Effect Transistor — semiconductor device whose insulated gate electrode controls current flow.

Fiber-optic: Relates to transmission of information as modulated light in tiny transparent fibers instead of copper wires.

Fibrin sealant: A biological tissue glue derived from human plasma. Sometimes referred to as fibrin glue, it contains the components necessary for the last step of coagulation: i) fibrinogen and other coagulation factors and ii) thrombin. It is used in surgical procedures to arrest bleeding and as an adjunct to wound healing.

Fibrin: A protein derived from FIBRINOGEN in the presence of THROMBIN, which forms part of the blood clot. [MeSH]

Fibrinogen: Plasma glycoprotein clotted by thrombin, composed of a dimer of three non- identical pairs of polypeptide chains (alpha, beta, gamma) held together by disulfide bonds. Fibrinogen clotting is a sol- gel change involving complex molecular arrangements: whereas fibrinogen is cleaved by thrombin to form polypeptides A and B, the proteolytic action of other enzymes yields different fibrinogen degradation products.

A plasma protein that, when activated by thrombin, becomes fibrin, the principle component of a blood clot. Plasma glycoprotein clotted by *thrombin*, composed of a dimer of three non- identical pairs of polypeptide chains (alpha, beta, gamma) held together by disulfide bonds. Fibrinogen clotting is a sol- gel change involving complex molecular arrangements: whereas fibrinogen is cleaved by thrombin to form *polypeptides* A and B, the proteolytic action of other enzymes yields different fibrinogen degradation products.

Fibroblast: Common cell type found in connective tissue. Secretes an extracellular matrix rich in collagen and other extracellular matrix macromolecules. Migrates and proliferates readily in wounded tissue and in tissue culture.

Fick's first law: The diffusion flux is proportional to the concentration gradient. This relationship is employed for steady-state diffusion situations. The time rate of change of concentration is proportional to the second derivative of concentration. This relationship is employed in non-steady-state diffusion situations.

Field-Effect Transistor (FET): Field-Effect Transistor — semiconductor device whose insulated gate electrode controls current flow.

Filler: An inert foreign substance added to a polymer to improve or modify its properties.

Firing: A high temperature heat treatment that increases the density and strength of a ceramic piece.

Fixative: Chemical reagent such as formaldehyde or osmium tetroxide used to preserve cells for microscropy. Samples treated with these reagents are said to be "fixed," and the process is called fixation.

Flagellum (plural flagella): Long, whiplike protrusion whose undulations drive a cell through a fluid medium. Eucaryotic flagella are longer versions of cilia; bacterial flagella are completely different, being smaller and simpler in construction.

Flat-band potential: *See* under Electrolyte/Insulator/ Silicon (EIS).

Flip-flop: binary device whose outputs change value only in response to an input pulse.

Fluidic Self Assembly: A novel technique for accurately assembling large numbers of very small devices. The small size, planarity, and accuracy of the assembly also result in very low parasitic interconnects, comparable to on die traces. This massively parallel assembly process combines the capability and flexibility of assembly with the cost effectiveness of integration.

Invented by Mr. Mark Hadley and was part of his Ph.D. dissertation while he was studying at University of California, Berkley. The FSA process became the foundation for the origins of a new company named Alien Technology Corporation. In the FSA process, specifically shaped semiconductor devices ranging in size from 10 microns to several hundred microns are suspended in liquid and flowed over a surface which has correspondingly shaped "holes" or receptors on it and into which the devices settle. The shape of the devices and of the holes is designed so that the devices fall easily into place and are selfaligning. Alien has successfully demonstrated the assembly of tens of thousands of devices in a single process step.

Fluorescein: Fluorescent dye that fluoresces green when illuminated with blue light or ultraviolet light.

Fluorescence: Luminescence (see also under "luminescence") which persists less than a second after the exciting cause has been removed. If the luminescence persists significantly longer it is called phosphorescence.

Fluorescent dye: Molecule that absorbs light at one

wavelength and responds by emitting light at another wavelength; the emitted light is of longer wavelength (and hence of lower energy) than the light absorbed.

FM: Frequency Modulation - information coding scheme in which the frequency of a steady wave is changed.

Focal contact (adhesion plaque): Small region on the surface of a fibroblast or other cell that is anchored to the extracellular matrix. The attachment is mediated by transmembrane proteins such as integrins, which are linked, through other proteins, to actin filaments in the cytoplasm.

Foglet: A mesoscale machine. A discreet component of utility fog. [J. S. Hall 1994]

Follicle cell: One of the cell types that surround a developing oocyte or egg.

Follower: A component in a cam system that is driven through a pattern of displacements as it rests against a moving contoured surface.

Forward bias: The conducting bias for a p-n junction rectifier that assures electron flow to the n side of the junction.

Fractal Mechatronic Universal Assembler: (or Fractal Assembler) is a machine that is capable of assembling any chemical from a generic descriptions of the properties required of the chemical. The machine comprises of test tube arrays and software linked to robotic cubes and sensor arrays to implement automated mixing and testing to conduct materials research activity. [FR] See Fractal Mechatronic Universal Assembler

Fractal Robots: AKA: Fractal Shape Shifting Robots and Programmable "Digital Matter", are programmable machines that can do unlimited tasks in the physical world, the world of matter. Load the right software and the same "machines" can vacuum the carpet, paint your car, or construct an office building and later, wash that building's windows. This is the beginning of "Digital Matter".

Fractal Shape Shifting Robots look like "Rubic's Cubes" that can "slide" over each other on command, changing and moving in any overall shape desired for a particular task. These cubes communicate with each other and share power through simple internal induction coils (or surface contacts in some models), have batteries, a small computer and various kinds of internal magnetic and electric inductive motors (depending on size) used to move over other cubes.

When sufficiently miniaturized (below 0.1mm) and fabricated using photolithography and E-Beam methods, the machines may exceed human manual dexterity and could then be programmed to assemble complex fractal aggregates or even to maintain the photolithographic and E-Beam equipment itself! The ultimate goal is self sustaining systems and "self-assembly" features that can drop cost dramatically and enable successive generations of robots exhibiting greater utility and value, to be built along the way. [Bill Spence]

Fractal: A mathematical construct that has a fractional dimension.

Free energy (G): A thermodynamic quantity that is a function of the enthalpy (H), the Kelvin temperature (T) and the entropy (S) of a system; G=H-TS. At equilibrium, the free energy is at a minimum. Under certain conditions the change in free energy for a process is equal to the maximum useful work. Energy that can be extracted from a system to drive reactions. Takes into account changes in both energy and entropy.

Free energy is a measure of the ability of a system to do work, such that a reduction in free energy could in principle yield an equivalent quantity of work. The Helmholtz free energy describes the free energy *within* a system; the Gibbs free energy does not.

Free Radical: A molecule containing an unpaired electron, typically highly unstable and reactive. Free radicals can damage the molecular machinery of biological systems, leading to cross-linking and mutation.

Free-energy change (DG): Change in the free energy during a reaction: the free energy of the product molecules minus the free energy of the starting molecules. A large negative value of DG indicates that the reaction has a strong tendency to occur.

Frequency response: Two relations between sets of inputs and outputs. One relates frequencies to the output-input amplitude ratio; the other relates frequencies to the phase difference between the output and input.

Frequency: Number of times per second that a quantity representing a signal, such as a voltage, changes state. Also, the number of waves (cycles) per second that pass a given point in space.

FUD: Fear, Uncertainty, Doubt.

Fullerene: A class of cage-like carbon compounds composed of fused, pentagonal and/or hexagonal sp^2 carbon rings. A new allotrope of carbon characterized by a closed cage structure consisting of an even number of three coordinate carbon atoms devoid of hydrogen atoms. This class was originally limited to closed-cage structures with twelve isolated five- membered rings, the rest being six- membered rings. [IUPAC Provisional Recommendations for the Nomenclature for the C_{60}-I_h and C_{70}-$D_{5h(6)}$ Fullerenes, 2001]

Fullerenes: Fullerenes are a molecular form of pure carbon discovered in 1985. They are cage-like structures of carbon atoms, the most abundant form produced is buckminsterfullerene (C60), with 60 carbon atoms arranged in a spherical structure. There are larger fullerenes containing from 70 to 500 carbon atoms. [Wid] What are fullerenes? A very good explanation, even though some of the translation is a bit rough. See also What are fullerenes?

Functional Group: Part of a molecule (group) with the potential to take part in a chemical reaction

Functional navigation: In medical nanorobotics, a form of nanorobotic navigation in which nanodevices seek to detect subtle variations in their environment, comparing sensor readings with target tissue/cell profiles and then congregating wherever a precisely defined set of preconditions exists.

G protein: One of a large family of heterotrimeric GTP-binding proteins that are important intermediaries in cell-signaling pathways. Usually activated by the binding of a hormone or other signaling ligand to a seven-pass transmembrane receptor protein.

G0 phase: (G-"zero" phase) State of withdrawal from the eucaroytic cell-division cycle by entry into a quiescent G1 phase; often seen in differentiated cells.

G1 phase: Gap 1 phase of the eucaryotic cell-division cycle, between the end of cytokinesis and the start of DNA synthesis.

G2 phase: Gap 2 phase of the eucaryotic cell-division cycle, between the end of DNA synthesis and the beginning of mitosis.

GAG (glycosaminoglycan): Long, linear, highly charged polysaccharide composed of a repeating pair of sugars, one of which is always an amino sugar. Mainly found covalently linked to a protein core in extracellular matrix proteoglycans. Examples include chondroitin sulfate, hyaluronic acid, and heparin.

Gain: The ratio of the amplitude of an output to input signal.

Galvanic corrosion: The preferential corrosion of the more chemically active of two metals electrically coupled and exposed to an electrolyte.

Gamete: Specialized haploid cell, either a sperm or an egg, serving for sexual reproduction.

Ganglion (plural ganglia): Cluster of nerve cells and associated glial cells located outside the central nervous system.

Ganglioside: Any glycolipid having one or more sialic acid residues in its structure. Found in the plasma membrane of eucaryotic cells and especially abundant in nerve cells.

Gap junction: Communicating cell-cell junction that allows ions and small molecules to pass from the cytoplasm of one cell to the cytoplasm of the next.

Gastrula: Animal embryo at an early stage of development where cells are invaginating to form the rudiment of a gut cavity. (From Greek *gaster*, belly.)

Gate: Circuit whose logical output variables are determined by its inputs. In digital logic, a component that can switch the state of an output dependent on the states of one or more inputs.

Gauss: The cgs unit used in measuring magnetic induction.

Gene: A given segment of the DNA molecule that contains the code for a specific protein.

Genegeneering: Genetic engineering.

General-purpose molecular manufacturing: A manufacturing technology that will find many applications across many segments of society. Its extreme flexibility, precision, high capacity, and low cost will cause rapid adoption almost everywhere, and therefore will have disruptive effects in many industries.

Genetic Algorithm: Any algorithm which seeks to solve a problem by considering numerous possibilities at once, ranking them according to some standard of fitness, and then combining ("breeding") the fittest in some way. In other words, any algorithm which imitates natural selection. [AS]

GENIE: An AI combined with an assembler or other universal constructor, programmed to build anything the owner wishes. Sometimes called a Santa Machine. This assumes a very high level of AI and nanotechnology. [AS]

Genomics: "The study of the full complement of genes that make up an organism."

Giant Magnetoresistance (GMR): It results from subtle electron-spin effects in ultra-thin 'multilayers' of magnetic materials, which cause huge changes in their electrical resistance when a magnetic field is applied. GMR is 200 times stronger than ordinary magnetoresistance. [See Spintronics and Giant Magneto Resistance] GMR enables sensing of significantly smaller magnetic fields, which in turn allows hard disk storage capacity to increase by a factor of 20.

Gibbs free energy: The Gibbs free energy is the Helmholtz free energy plus the product of the system volume and the external pressure. Changes in the Gibbs free energy at a constant pressure thus include work done against external pressure as a system undergoes volumetric changes. This proves convenient for describing equilibria in gases and liquids at a constant pressure (e.g., at one atmosphere), but is of little use in describing machine-phase chemical processes. Changes in the Gibbs free energy caused by a change in the applied pressure (at constant volume) have no direct physical significance. (See also enthalpy.)

Glass transition temperature (T_g): The temperature at which, upon cooling, a noncrystalline ceramic or polymer transforms from a supercooled liquid to a rigid glass.

Glass: An amorphous solid obtained when silica is mixed with other compounds, heated above its melting point, and then cooled rapidly.

GNR technologies: (Genetic Engineering, Nanotechnology, and Robotics)

Golden Goo: Another member of the grey goo family of nanotechnology disaster scenarios. The idea is to use nanomachines to filter gold from seawater. If this process got out of control we would get piles of golden goo (the "Wizard's Apprentice Problem").

This scenario demonstrates the need of keeping populations of self-replicating machines under control; it is much more likely than grey goo, but also more manageable. [AS - Originated on sci.nanotech 1996]

Grain boundary: The interface separating two adjoining grains having different crystallographic orientations.

Grain growth: The increase in average grain size of a polycrystalline material: for most materials, an elevated temperature heat treatment is necessary.

Grain size: The average grain diameter as determined from a random cross section.

Gray Goo or Grey Goo - destructive nanobots [AKA: "gray dust"]: Opposite of Blue Goo. See Star Trek scenario. Vast legions of *destructive* nanites. Typically, created by accident. Left unchecked, they will basically convert everthing they contact into more of themselves, or consume and digest it for energy. Either way, its pretty much bad news. Also - Self-replicating (von Neumann) nanomachines spreading uncontrolably, building copies of themselves using all available material. This is a commonly mentioned nanotechnology disaster scenario, although it is rather unlikely due to energy constraints and elemental abundances. More probable disaster scenarios are the green goo, golden goo, red goo, khaki goo scenarios. As a protection blue goo has been proposed.

Gray goo problem: Nanotech Now's Goo and Paste Glossary. The following quote was taken from Eric Drexler's book Engines of creation.

The early transistorized computers soon beat the most advanced vacuum-tube computers because they were based on superior devices. For the same reason, early assembler-based replicators could beat the most advanced modern organisms. "Plants" with "leaves" no more efficient than today's solar cells could out-compete real plants, crowding the biosphere with an inedible foliage. Tough, omnivorous "bacteria" could out-compete real bacteria: they could spread like blowing pollen, replicate swiftly, and reduce the biosphere to dust in a matter of days. Dangerous replicators could easily be too tough, small, and rapidly spreading to stop - at least if we made no preparation. We have trouble enough controlling viruses and fruit flies.

Among the cognoscenti of nanotechnology, this threat has become known as the "gray goo problem." Though masses of uncontrolled replicators need not be gray or gooey, the term "gray goo" emphasizes that replicators able to obliterate life might be less inspiring than a single species of crabgrass. They might be "superior" in an evolutionary sense, but this need not make them valuable. We have evolved to love a world rich in living things, ideas, and diversity, so there is no reason to value gray goo merely because it could spread. Indeed, if we prevent it we will thereby prove our evolutionary superiority. The gray goo threat makes one thing perfectly clear: we cannot afford certain kinds of accidents with replicating assemblers.

Gray goo: The name given to free-range self-replicating miniature machines that could, in theory, run out of control and cause severe damage to the biosphere. The actual threat is generally overrated, as we explain here.

Green ceramic body: A ceramic piece, formed as a

particulate aggregate, that has been dried but not fired.

Green Goo: Nanomachines or bio-engineered organisms used for population control of humans, either by governments or eco-terrorist groups. Would most probably work by sterilizing people through otherwise harmless infections. See Nick Szabo's essay Green Goo — Life in the Era of Humane Genocide. [AS] "A scary concept dreamed up by Erik K Drexler whereby tiny *assemblers*, or molecular machines, that are capable of making copies of themselves, are let loose and proceed to replicate uncontrollably, consuming everything in their path and turning it into a grey goo."

Ground state: The lowest-energy state of a system. The electronic ground state of a system cannot reduce its energy by an electronic transition, but may contain vibrational energy (kinetic and potential energy associated with the motions and positions of its atoms); extended systems at ordinary temperatures are always vibrationally excited, and so "ground state" is often taken to mean "electronic ground state."

Ground: To make electrical connection to the earth or to the chassis of a device (verb); the connection point so used (noun).

Group velocity: In wave propagation, the speed of the waveform (e.g., of a peak) can be different from the speed of a group of waves (e.g., of a set of ripples in water). The latter is the *group velocity*, and is the speed of propagation of information and wave energy. The waveform speed is the *phase velocity*.

Group: A set of linked atoms in a molecule; a defined

substructure. Typically, a set that is usefully regarded as a unit in chemical reactions of interest.

GUI: Graphical User Interface — hardware, software, and firmware that produces the display on modern personal computers.

Guy Fawkes Scenario: If nanotechnology becomes widely available, it might become trivial for anyone to committ acts of terrorism (such as making nanomachines build a large amount of explosives under government buildings a la Guy Fawkes). This would either force strict control over nanotechnology (hard) or a decentralized mode of organization.

substances. Typically, each facet is usually regarded as a unit in chemical reactions or interactions.

H

Hair cell: Specialized sensory epithelial cell in the ear with bundles of giant microvilli (stereocilia) protruding from its apical surface. Sound vibrations tilt the stereocilia, evoking an electrical change in the hair cell, which thus acts as a sound detector.

Hall effect: The phenomenon whereby a force is applied to a moving electron or hole by a magnetic field that is applied perpendicular to the direction of motion. The force direction is perpendicular to both the magnetic field and the particle motion directions.

Haploid: Having only one set of chromosomes, as in a sperm cell or a bacterium, as distinct from diploid (having two sets of chromosomes).

Hardness: The measure of a material's resistance to deformation by surface indentation or by abrasion. There are various scales in use to express hardness. The Mohs scale is qualitative and somewhat arbitrary and ranges from 1 on the soft end for talc to 10 for diamond. Quantitative scales are the Rockwell (HR), Brinell (indicated by HB), Knoop (HK) and Vickers (HV). Knoop and Vickers are referred to as microhardness testing methods on the basis of load and indenter size.

Harmonic oscillator: A system in which a mass is subject to a linear restoring force, like an ideal spring. A harmonic oscillator vibrates at a fixed frequency, independent of amplitude.

Heat capacity (C_v at constant volume and C_p at constant pressure): The quantity of heat required to produce a unit temperature rise per mole of material. The ratio of the heat input to the temperature increase in a system. Note that this definition does not imply that a system *contains* heat, despite the name *heat capacity*.

Heat shock protein (stress-response protein): Protein synthesized in response to an elevated temperature or other stressful treatment; usually helps the cell to survive the stress.

Heat: As defined in thermodynamics, heat is the energy that flows between two systems as a result of temperature differences (a system *contains* neither heat nor work, but can *produc*e heat or *do* work). Heat thus differs from thermal energy.

Heisenberg Uncertainty Principle: A quantum-mechanical principle with the consequence that the position and momentum of an object cannot be precisely determined. The Heisenberg principle helps determine the size of electron clouds, and hence the size of atoms.

HeLa cell: Line of human epithelial cells that grows vigorously in culture. Derived from a human cervical carcinoma.

Helmholtz free energy: The internal energy of a system minus the product of its entropy and temperature; see free energy.

Heme: An iron complex. Cyclic organic molecule containing an iron atom that carries oxygen in hemoglobin and carries an electron in cytochromes.

Hemidesmosome: Specialized cell junction between an epithelial cell and the underlying basal lamina.

Hemoglobin: A biomolecule composed of four myoglobine-like units (proteins plus heme) that can bind and transport four oxygen molecules in the blood.

The major protein in red blood cells that associates with O_2 in the lungs by means of a bound heme group.

Hemopoiesis (hematopoiesis): Generation of blood cells, mainly in the bone marrow.

HEMT: High electron mobility transistor.

Henry (H): Unit of inductance (see"inductance"). One henry (H) is the inductance of a closed circuit in which an electromotive force of 1 volt is produced when the electric current in the circuit varies uniformly at the rate of 1 ampere per second.

Henry's law: The amount of gas dissolved in a solution is directly proportional to the pressure of the gas above the solution.

Heterocaryon: Cell with two or more nuclei produced by the fusion of two or more different cells.

Heterochromatin: Region of a chromosome that remains unusually condensed and transcriptionally inactive during interphase.

Heterodimer: Protein complex composed of two different polypeptide chains.

Heteronuclear: Consisting of different elements. [TNTWeekly]

Heterozygote: Diploid cell or individual having two different alleles of a specified gene.

Heuristics: Rules of thumb used to guide one in the direction of probable solutions to a problem.

Highest Occupied Molecular Orbital (HOMO): The highest energy molecular orbital of an atom or molecule that contains an electron.

Histonatation: In medical nanorobotics, locomotion (swimming) through tissues by a nanorobot.

Histone: A protein present in the eukaryotic nucleus that is bound to the DNA at regularly spaced intervals.

Homogeneous and Heterogeneous Assays: A homogeneous assay does not require a separation step to remove free antigen from bound antigen and relies upon the fact that the function of the label is modified upon binding, leading to a change in signal intensity. Because of high background signal a heterogeneous approach incorporating a separation step of bound and unbound makes the detection limit lower, approaching the values obtained by RIA. The homogeneous assay is less technically demanding.

Human Genome Project: A research initiative that has mapped the entire human genome.

Human tissue engineering No universally agreed definition exists. But a Commission working hypothesis suggests a human tissue engineered product means any autologous (emanating from the patient himself) or allogeneic (coming from another human being) product which: contains, consists of, or results in engineered human cells or tissues; and has properties for, or is presented as having properties for, the regeneration, repair or replacement of tissue, where the new tissue or cells, in whole or in part, are structurally and functionally analogous to the original tissue that is being regenerated, repaired or replaced.

Hydrocarbon: A molecule consisting only of H and C.

Hydrogen Bond: A bond formed between hydrogen and two other atoms. A hydrogen atom covalently bound to an electronegative atom (e.g., nitrogen, oxygen) has a significant positive charge and can form a weak bond to another electronegative atom; this is termed a hydrogen bond.

Hydrophobic force: Water molecules are linked by a network of hydrogen bonds. A nonpolar, nonwetting, surface (e.g., wax) cannot form hydrogen bonds. To form their full complement of hydrogen bonds, the nearby water molecules must form a more orderly (hence lower entropy) network. This both increases free energy and causes forces that tend to draw hydrophobic surfaces together across distances of several nanometers.

Hydrophobic: Literally - water fearing, from the Greek *hydro* - "water" and *phobo* - "fear".

Hypertext: A computer-based system for linking text and other information with cross-references, making access fast and criticisms easy to publish and find.

Hypsithermal limit: The maximum amount of energy that may be released at Earth's surface, as a result of human technological activities, without significantly altering the natural global energy balance; estimated as 10^13-10^15 watts.

Hysteresis: The difference in the output when a specific input value is approached first with an increasing and then with a decreasing input. This phenomenon occurs in ferroelectric materials and results in irreversible loss of energy through heat dissipation.

I

I/O: Input/output information transfer between computer and peripherals such as keyboard or printer.

IA: Intelligence Amplification: Technologies seeking to increase the cognitive abilities of people.

IC: See"integrated circuit".

Imaging Contrast Agent: A molecule or molecular complex that increases the intensity of the signal detected by an imaging technique, including MRI and ultrasound. An MRI contrast agent, for example, might contain gadolinium attached to a targeting antibody. The antibody would bind to a specific target—a metastatic melanoma cell, for example – while the gadolinium would increase the magnetic signal detected by the MRI scanner.

Imaging Contrast Agent: A molecule or molecular complex that increases the intensity of the signal detected by an imaging technique, including MRI and ultrasound. An MRI contrast agent, for example, might contain gadolinium attached to a targeting antibody. The antibody would bind to a specific target—a metastatic melanoma cell, for example – while the gadolinium would increase the magnetic signal detected by the MRI scanner.

Immortalization: Production of a cell line capable of an unlimited number of cell divisions. Can be the result of a chemical or viral transformation or of fusion with cells of a tumor line.

Immune Machines: Medical nanomachines designed for internal use, especially in the bloodstream and digestive tract, able to identify and disable intruders such as bacteria and viruses.

Immune response: Response made by the immune system of a vertebrate when a foreign substance or microorganism enters its body.

Immune system: Population of lymphocytes and other white blood cells in the vertebrate body that defends it against infection.

Immunoglobulin (Ig): An antibody molecule. Higher vertebrates have five classes of immunoglobulin - IgA, IgD, IgE, IgG, and IgM - each with a different role in the immune response.

Immunoglobulin like (Ig-like) domain: Characteristic protein domain of about 100 amino acids that is found in antibody molecules and in many other proteins that form the Ig superfamily.

IMP: Electronic implant, especially in the brain. [Ron Hale Evans]

Impedance: The complex ratio of a force-like quantity (force, pressure, voltage, temperature, or electric field) to a corresponding related velocity-like quantity (velocity, volume velocity, current, heat flow, or magnetic field strength).

In cyto: Within a biological cell.

In nucleo: Within the nucleus of a cell.

In nucleo: Within the nucleus of a cell.

In sanguo: Within the bloodstream.

***In situ* hybridization:** Technique in which a single-stranded RNA or DNA probe is used to locate a gene or an mRNA molecule in a cell or tissue. (See also hybridization.)

***In vitro*:** Term used by biochemists to describe a process taking place in an isolated cell-free extract. Also used by cell biologists to refer to cells growing in culture (*in vitro*), as opposed to in an organism (*in vivo*). (Latin for "in glass.")

***In vivo*:** In an intact cell or organism. (Latin for "in life.")

Index of refraction (n): The ratio of the velocity of light in a vacuum to the velocity in some medium.

Inductance [in Henry, H]: That property of an electric circuit which tends to oppose change in current in the circuit. One henry (H) is the inductance of a closed circuit in which an electromotive force of 1 volt is produced when the electric current in the circuit varies uniformly at the rate of 1 ampere per second.

Induction (embryonic): Change in the developmental fate of one tissue caused by an interaction with another tissue.

Inductor: Energy storage circuit component consisting of a coil of wire and possibly a magnetic material.

Inflammatory response: Local response of a tissue to injury or infection. Caused by invasion of white blood cells, which release various local mediators such as histamine.

Infrared: Invisible electromagnetic radiation having a longer wavelength, and lower frequency, than visible red light.

Inhibitor: A chemical substance that, when added in relatively low concentrations, slows down a chemical reaction.

Inline Universities: (As opposed to online universities), nanocomputer implants serving to increase intelligence and education of their owners, essentially turning them into walking universities.

Inmessaging: In medical nanorobotics, conveyance of information from a source external to the human body, or external to working nanodevices, to a receiver located inside the human body.

Insertion point (in lithography context): Adaptation of a new lithography technique is referred to as the insertion point of that technique.

Insulator: A substance, object or material that does not conduct electricity. Material that conducts electricity very poorly.

Integrated Circuit (IC): An electronic circuit consisting of many interconnected devices on one piece of semiconductor, typically into 10 millimeters on a side. ICs are the major building blocks of today's computers. Semiconductor circuit, typically on a very small silicon chip, containing microfabricated transistors, diodes, resistors, capacitors, etc.

Integrated microdevices IM: DARPA/MTO wishes to capitalize on the investments in the manufacture, design and test of **semiconductor** integrated circuits by applying them to the emerging mixed technology systems area. The tight integration of microdevices in mixed technology systems, however, requires more than just an electronic domain analysis to understand and optimize the functionality of the design. Issues such as coupled energy domain simulation; three dimensional shape analysis before, and possibly after, integration; and mixed technology interconnect design and analysis need to be addressed as part of the design process. Key to enabling mixed technology systems is the development of a design environment that supports design and manufacture based upon many available mixed technology and electronic building blocks. In addition, design trade-offs, optimizations and synthesis need to be explored from an overall systems perspective in a mixed domain design and layout environment. The ultimate goal of the Composite CAD program is to provide a validated design environment that enables the reliable and predictable design of mixed technology systems. [Heather Dussault Microsystems Technology Office DARPA "Design for Mixed Technology Integration (Composite CAD)" 2001]

A new term, integrated microdevices (or IM's), is being introduced to identify this new type of monolithic system. DARPA seeks innovative proposals in the following areas: I. Process and shape based analysis for the process and device engineer, II. Circuit and physical layout synthesis and verification environments for the systems designer, and III. Applications which utilize the resulting design technology on integrated microdevices (IM's). The primary focus is on linking under- development shape based analysis tools into a functional circuit and physical layout environment. The circuit and layout environment could be enhanced from standard mixed signal design tool set but needs to allow the design and incorporation of mixed technology devices such as microelectromechanical (MEMS) resonators, optoelectronic transmitters and receivers, and microfluidic valves, channels and chambers. [Microsystems Technology Office, Defense Advanced Research Projects Agency, US Commerce Business Daily Feb. 11, 1997]

Intelligent Agent: Aka "software agent". Software that can do things without supervision, because it knows your patterns, history, preferences, likes, dislikes, and so forth. You want to take a vacation - it knows that you really enjoyed that trip to Hawaii, and that you prefer to fly at night, 1st class. It also knows that the bungalow you rented last time was marked as being 5-star, and worth a re-visit. Your IA then collates all your parameters, searches the internet for flights, car rentals, restaurant reservations, and lodgings, and schedules everything for you, with options on the side. No more travel agent - you have a software agent to handle things! Many experts agree that by 2010 we will each have one, and that they will greatly reduce our daily load of trivial and redundant tasks. See Is There an Intelligent Agent in Your Future?.

Interagency Working Group on Nanoscience, Engineering and Technology IWGN, National Science and Technology Council NSTC, US

Interface: In physical terms, an interface is the boundry between two phases, for instance between a solid and liquid or between a liquid and gas.

Intermolecular: Between more than one molecule. Describes an interaction (e.g., a chemical reaction) between different molecules.

Internal energy: The sum of the kinetic and potential energies (including electromagnetic field energies) of the particles that make up a system.

Internet: Worldwide digital communication network in which packets of information travel between senders and recipients.

Interstitial diffusion: a diffusion mechanism that causes atomic motion from interstitial site to interstitial site.

Intramolecular: Describes an interaction (e.g., a chemical reaction) within a single molecule. Intramolecular interactions between widely separated parts of a molecule resemble intermolecular interactions in most respects. (2) Within a single molecule.

Intrinsic: Characterizes pure undoped semiconductor; electrical conductivity depends only on temperature and the band gap energy.

Ion: An atom or molecule with a net charge. **(2)** An atom with more or fewer electrons than those needed to cancel the electronic charge of the nucleus. An ion is an atom with a net electric charge.

Ionic bond: A chemical bond resulting chiefly from the electrostatic attraction between positive and negative ions. A coulombic interatomic bond existing between two adjacent and oppositely charged ions.

Ionophore: A macro-organic molecule capable of specifically solubilizing an inorganic ion of suitable size in organic mediums.

ISE (Ion selective electrode): Ions in solution are quantified by measuring the change in voltage (i.e. potentiometric) resulting from the distribution of ions (by ion exchange controlled by the ion exchange current io) between a sensing membrane (the ion selective membrane) and the solution. This potential is measured at zero current with respect to a reference electrode which is also in contact with the solution. The potential measured is proportional to the logarithm of the analyte concentration. The oldest and best known ISE is the pH sensor based on a glass membrane. More recently, polymeric membranes have been formed incorporating ionophores (see ionophore) rendering the membrane specific to certain ions only.

ISFET: Ion Sensitive Field Effect Transistor: a logical extension of ISE's. They can be conceptualized by imagining that the lead from an ion-selective electrode, attached via a cable to a FET in the high impedance input stage of a voltmeter, is made shorter until no lead exists and the selective membrane is attached directly to the FET. For an ISFET, the property measured is the lateral conductivity between two opposing doped regions (the source and drain) surrounding the active area. The underlying change is a change in flat-band voltage.

Isoelectronic: Two molecules are described as isoelectronic if they have the same number of valence electrons in similar orbitals, although they may differ in their distribution of nuclear charges (e.g., H-CN and $H-N^+C^-$. KineticPertaining to the rates of chemical reactions. A fast reaction is said to have fast kinetics; if the balance of products in a reaction is controlled by reaction rates rather than by thermodynamic equilibria, the reaction is said to be kinetically controlled.

Isomorphous: Having the same structure. In the phase diagram sense, isomorphicity means having the same crystal structure or complete solid solubility for all compositions.

Isothermal: At a constant temperature. In an isothermal process heat is, if necessary, supplied or removed from the system at just the right rate to maintain constant temperature.

Isotropic: Having identical values of a property in all crystallographic directions.

J

Joule: The joule (pronounced DJOOL) is the standard unit of *energy* in electronics and general scientific applications. One joule is defined as the amount of energy exerted when a force of one is applied over a displacement of one meter. One joule is the equivalent of one watt of *power* radiated or dissipated for one second. In some applications, the British thermal unit (Btu) is used to express energy. One Btu is equivalent to approximately 1055 joules. (2) Standard unit of energy in the meter-kilogram-screened system. One joule is the energy delivered in one second by a one-watt power source. Approximately equal to 0.24 calories.

Julia Set: Complex fractal derived from iterating a set of non-diverging values. There is an infinite number of possible Julia sets, each of which corresponds to a particular complex number that appears as a constant in the iterative procedure. A number of hyperturing clades and aioids find experimenting with Julia Sets aesthetically pleasing. All Julia sets are related to, and included in, the Mandelbrot set.

Junna: Minor Negentropy Alliance world. During the period from 3267 to 3421 was center for the Institute of Baseline Psychotypology. Suffered during the Version War and became an MPA garrison. Since the age of separative empires the locals have drifted increasingly towards Zoeticism

Jupiter-Brain: A posthuman being of extremely high computational power and size. This is the archetypal concentrated intelligence. The term originated due to an idea by Keith Henson that nanomachines could be used to turn the mass of Jupiter into computers running an upgraded version of himself.

K

Karyotype: Full set of chromosomes of a cell arranged with respect to size, shape, and number.

Keratin (cytokeratin): Member of the family of proteins that form keratin intermediate filaments, mainly in epithelial cells. Some specialized keratins are found in hair, nails, and feathers.

Ketone: Organic molecule containing a carbonyl group linked to two alkyl groups.

Kevlar (TM): A synthetic fiber made by E. I. du Pont de Nemours & Co., Inc. Stronger than most steels, Kevlar is among the strongest commercially available materials and is used in aerospace construction, bulletproof vests, and other applications requiring a high strength-to-weight ratio.

Khaki Goo: Military nanotechnology; see grey goo.

Kilo-: Prefix denoting 1E3.

Kilobyte (kB): 2^{10} (= 1024, or about one thousand) bytes of information.

Kilocalorie (kcal): Unit of heat energy equal to 1000 calories. Often used to express the energy content of food or molecules: bond strengths, for example, are measured in kcal/mole. An alternative unit in wide use is the kilojoule, equal to 0.24 kcal.

Kilohertz (kHz): One thousand cycles per second (see also"frequency").

Kilojoule: Standard unit of energy equal to 1000 joules, or 0.24 kilocalories.

Kinesin: One type of motor protein that uses the energy of ATP hydrolysis to move along a microtubule.

Kinetic energy: Energy resulting from the motion of masses.

Kinetic molecular theory: A model that assumes that an ideal gas is composed of tiny particles (molecules) in constant motion.

Kinetochore: Complex structure formed from proteins on a mitotic chromosome to which microtubules attach and which plays an active part in the movement of chromosomes to the pole. The kinetochore forms on the part of the chromosome known as the centromere.

Knowbots: Knowledge robots, first developed Vinton G. Cref and Robert E. Kahn for National Research Initiatives. Knowbots are programmed by users to scan networks for various kinds of related information, regardless of the language or form in which it expressed. "Knowbots support parallel computations at different sites. They communicate with one another, and with various servers in the network and with users."

L

Label or marker: A problem endemic in immunoassays is the absence of a chemical signal created by the antibody-antigen binding, in contrast with an enzyme-substrate binding reaction which produces a chemical reaction product. As a result of this absence, the use of a label or marker is usually required to detect the bound antibody-antigen complex. Several markers have been established for use in immunoassays. Examples of such markers are:

- Particles (e.g. latex, gold particles, erythrocytes)
- Metal and dye sols (e.g. Au, Palanil®; Luminous Red G)
- Chemiluminescent and bioluminescent compounds (e.g. Luciferase/luciferins, Luminol and derivatives, Acridinium esters, Peroxidase)
- Electrochemical active species (ions, redox species, ionophores)
- Fluorophores (e.g. Dansyl chloride DANS, Rare earth metal chelates, Umbelliferones)
- Chromophores
- Enzymes (e.g. Alkaline phosphatase, â-D-Galactosidase, Peroxidase), substrates, cofactors

- Liposomes
- Iodine-125, tritium, ^{14}C, ^{75}Se, ^{57}Co.

Label: Chemical group or radioactive atom added to a molecule in order to follow its progress through a biochemical reaction or to locate it spatially. Also, as a verb, to add such a group or atom to a cell or molecule.

Lagging strand: One of the two newly made strands of DNA found at a replication fork. The lagging strand is made in discontinuous lengths that are later joined covalently.

Lambda bacteriophage (l bacteriophage): Virus that infects *E. coli;* widely used as a DNA cloning vector.

Lamin: Extracellular matrix protein found in basal laminae.

Lamins: Intermediate filament proteins that form the fibrous matrix (nuclear lamina) on the inner surface of the nuclear envelope.

Langmuir-Blodgett (LB) Film: A film of organic material (often surfactant molecules) assembled at the liquid-gas interface and subsequently transferred onto a solid substrate.

Langmuir-Blodgett: The name of a nanofabrication technique used to create ultrathin films (monolayers and isolated molecular layers), the end result of which is called a "Langmuir-Blodgett film". See *Nanotechnology and Monolayers* for more information. More and more.

Laser trimming: A method for adjusting the value of thin- or thick-film resistors by using a computer-controlled laser system.

Laser tweezers: Laser tweezers and related approaches

are important for manipulation and isolation of subcellular *organelles* and structural components. [National Center for Research Resources "Integrated Genomics Technologies Workshop Report" Jan 1999]

Laser: Light Amplification by the Stimulated Emission of Radiation — quantum device that produces coherent light.

LCD (Liquid Crystal Display): LCD is the predominant technology used in flat panel displays. The principle that makes the display work is this: A crystalís alignment can be altered with an electric current. If the crystal is lined up one way ñ it will allow the light waves to pass through a polarized filter, but if the electric current alters the crystalís alignment, it will guide light so that the polarized filter blocks the light. By densely packing red, blue and green light emitting crystals next to each other on a sheet (ìcalled a substrateî), one can create a full color display. The great thing about LCD is that the crystals can be packed together closely, allowing for a higher-resolution, finer-detail display. The con is that LCDs are somewhat fragile, require a lot of power and are relatively less bright. Liquid Crystal Display — display device employing light source and electrically alterable optically active thin film.

Leading strand: One of the two newly made strands of DNA found at a replication fork. The leading strand is made by continuous synthesis in the 5′-to-3′ direction.

Leakage: The loss of all or parts of a useful agent, as of the electric current that flows through an insulator or the magnetic flux that passes outside useful flux circuits.

Lectin: Protein that binds tightly to a specific sugar. Abundant lectins derived from plant seeds are often used as affinity reagents to purify glycoproteins or to detect them on the surface of cells.

LED: Light-Emitting diode — semiconducting diode that produces visible or infrared radiation.

LEDs (Light Emitting Diodes): LEDs work on a completely different concept. Traditionally LEDs are created from two semiconductors. By running current in one direction across the semiconductor the LED emits light of a particular frequency (hence a particular color) depending on the physical characteristics of the semiconductor used. The semiconductor is covered with a piece of plastic that focuses the light and increases the brightness. These semiconductors are very durable, there is no filament, they donít require much power, theyíre brighter and they last a long time. By densely packing red, blue and green LEDs next to each other on a substrate one can create a display.

The disadvantage of LEDs is that they are much larger ñ therefore the resolution is not nearly as good as LCD displays. Thatís why most LED displays are large, outdoor displays, not smaller devices, like monitors.

Leucine zipper: Structural motif seen in many DNA-binding proteins in which two a helices from separate proteins are joined together in a coiled-coil, forming a protein dimer.

Lewis acid: An electron-pair acceptor.

Lewis base: An electron-pair donor.

Life (lifetime): The length of time the sensor can be used before its performance changes.

Life Science Applications of Deep Red T2-MP EviTag Quantum Dots: The new T2-MP EviTags are available in the deep red wavelengths used in life science research – Macoun Red (660nm) and Jonamac Red (680nm) – with carboxyl, amine, biotin and non-functionalized surfaces.

Having a biologically compatible quantum dot label like the T2-MP EviTags allows further exploration and development of in vivo assays, following cells inside living organism to get a new understanding of life. In particular, it offers the possibility to create new classes of assays for use where current testing materials could produce cytotoxicity or cell death. Potential applications include live cell and even whole animal imaging, blood cancer assays, and numerous other applications.

Ligand: In protein chemistry, a small molecule that is (or can be) bound by a larger molecule is termed a ligand. In organometallic chemistry, a moiety bonded to a central metal atom is also termed a ligand; the latter definition is more common in general chemistry.

Light Emitting Diodes (LEDs): Light Emitting Diodes work on a completely different concept. Traditionally LEDs are created from two semiconductors. By running current in one direction across the semiconductor the LED emits light of a particular frequency (hence a particular color) depending on the physical characteristics of the semiconductor used. The semiconductor is covered with a piece of plastic that focuses the light and increases the brightness. These semiconductors are very durable, there is no filament, they donít require much power, theyíre brighter and they last a long time. By densely packing red, blue and green LEDs next to each other on a substrate one can create a display.

Lightsail: A spacecraft propulsion system that gains thrust from the pressure of light striking a thin metal film.

Limit of detection: The smallest measurable input. This differs from resolution, which defines the smallest measurable change in input. For a temperature measurement, this would provide an indication of the lowest temperature a sensor could generate an output in response to.

Limited Assembler: An assembler with built-in limits that constrain its use (for example, to make hazardous uses difficult or impossible, or to build just one thing). Assembler capable of making only certain products; faster, more efficient, and less liable to abuse than a general-purpose assembler.

Linde Scenario: A scenario for indefinite survival of intelligent life. It assumes it is possible to either create basement universes connected to the original universe with a wormhole or the existence of other cosmological domains. Intelligent life continually migrates to the new domains as the old grow too entropic to sustain life. [AS/Mitch Porter, 1997. The name refers to Linde's chaotic inflation cosmology, where new universes are continually spawned.]

Linear coefficient of thermal expansion: *See* thermal expansion coefficient, linear.

Linear: Aside from its geometric meaning, *linear* describes systems in which an output is directly proportional to an input. In particular, a linear elastic system is one in which the internal displacements are (at equilibrium) directly proportional to applied forces.

Linearity: The degree to which the calibration curve of a device conforms to a straight line.

Lipids: Water-insoluble substances than can be extracted from cells by nonpolar organic solvents.

Liposome: A type of nanoparticle made of lipids, or fat molecules, surrounding a water core. Liposomes, several of which are widely used to treat infectious diseases and cancer, were the first type of nanoparticle to be used to create therapeutic agents with novel characteristics.

Liposome: A type of nanoparticle made of lipids, or fat molecules, surrounding a water core. Liposomes, several of which are widely used to treat infectious diseases and cancer, were the first type of nanoparticle to be used to create therapeutic agents with novel characteristics.

Liquid Crystal Display: LCD is the predominant technology used in flat panel displays. The principle that makes the display work is this: A crystalís alignment can be altered with an electric current. If the crystal is lined up one way ñ it will allow the light waves to pass through a polarized filter, but if the electric current alters the crystalís alignment, it will guide light so that the polarized filter blocks the light. By densely packing red, blue and green light emitting crystals next to each other on a sheet (ìcalled a substrateî), one can create a full color display. The great thing about LCD is that the crystals can be packed together closely, allowing for a higher-resolution, finer-detail display. The con is that LCDs are somewhat fragile, require a lot of power and are relatively less bright. Liquid Crystal Display — display device employing light source and electrically alterable optically active thin film.

Lithography: A process for creating chemical patterns on a surface.

LMNT: An abbreviation for *limited* molecular nanotechnology; a narrowly specified type of MNT, using only diamondoid reactions; much easier to achieve than general MNT, but with nearly equivalent appeal and impact.

Lofstrom Loop: An beanstalk-like megaconstruction based on a stream of magnetically accelerated bars linked together. The stream is sent into space, where a station rides it using magnetic hooks, redirects it horizontally to another station, which sends it downwards to a receiving station on the ground. From this station the stream is then sent back to the launch station (a purely vertical version is called a space fountain).

This structure would contain a large amount of kinetic energy but could be built gradually and would only require enough energy to compensate for losses when finished. Elevators could be run along the streams, and geostationary installations could be placed along the horizontal top. [Named after Keith Lofstrom, who did the first detailed calculations on it in: Lofstrom, Keith H., "The launch loop — a low cost Earth-to-high orbit launch system," AIAA Paper 85-1368, 1985].

London dispersion force: An attractive force caused by quantum-mechanical electron correlation. For example, a neutral spherical molecule (such as a single argon atom) has no charge and produces no external electric field, yet a pair of molecules has a distribution of electron configurations weighted toward those with lesser electron-electron repulsions; this creates a small net attraction. (2) An attractive force between atoms or molecules caused by the numerous transient dipoles resulting from electronic superposition.

Lone pair: Two valence electrons of an atom that share an orbital but do not participate in a bond.

Low-dimension Structures: Quantum wells, quantum wire and quantum dots.

Lowest Unocupied Molecular Orbital (LUMO): The lowest energy molecular orbital of an atom or molecule that does not contain an electron.

Luminescence: Defined as the mission of light from a substance in an electronically excited state. Depending on whether the excited state is singlet or triplet, the emission is called fluorescence (less than one second decay) or phosphorescence (longer than 1 second decay). Depending on the source, molecules get the needed extra energy from different types of luminescence are distinguished: radioluminescence, photoluminescence (in the same category are fluorescence and phosphorescence), chemiluminescence and bioluminescence, electrochemiluminescence, sonochemi-luminescence and thermoluminescence.

M

Machine-phase chemistry: The chemistry of systems in which all potentially reactive moieties follow controlled trajectories (e.g., guided by molecular machines working in vacuum).

Macroscale: Larger than nanoscale; often implies a design that humans can directly interact with; too large to be built by a single assembler (one cubic micron of diamond contains 176 billion atoms).

Macrosensing: In medical nanorobotics, the detection of global somatic states (inside the human body) and extrasomatic states (sensory data originating outside of the human body) by in vivo nanorobots.

Magnetic field strength (designated by H) [A/m]: Magnetic field produced by a current, independent of the presence of magnetic material. The units of H are ampere-turns per meter, or just amperes per meter.

Magnetic flux density or magnetic induction (designated by B): The magnetic field produced in a substance by an external magnetic field. The units of B are tesla (T). One tesla is the magnetic flux density given by a magnetic flux of 1 weber per square meter. One weber is a magnetic flux that, linking a circuit of 1 turn, would produce in it an electromotive force of 1 volt if it were reduced to zero at a uniform rate in 1 second. Both B and H are field vectors. One henry (H) is the inductance of a closed circuit in which an electromotive force of 1 volt is produced when the electric current in the circuit varies uniformly at the rate of 1 ampere per second. The magnetic field strength and flux density are related according to: $B = \mu H$, where μ is the permeability (see under permeability).

Magnetic Force Microscopy (MFM): A method for observing local magnetic fields near a surface by scanning the surface with a magnetic probe.

Magnetic susceptibility ($\div_m$): The proportionality constant between the magnetization M (see under"magnetization") and the magnetic field strength H. The magnetic susceptibility is unitless.

Magnetization (M): The total magnetic moment per unit volume of material. Also, a measure of the contribution to the magnetic flux by some material within an H field. The magnitude of M is proportional to the applied field as: $M = \div_m \times H$, with $\div_m$ the magnetic susceptibility.

Magnetostrictive material: A material that changes dimension in the presence of a magnetic field or generates a magnetic field when mechanically deformed.

Manufacturing: The ability to make products, in this case ranging from clothing, to electronics, to medical devices, to books, to building materials, and much more.

Martensite: A metastable iron phase supersaturated in carbon that is the product of a diffusionless (athermal) transformation from austenite.

Mask: Pattern on glass, like a photographic negative, for producing integrated-circuit elements on semiconductor wafer.

Mass to Charge Ratio: A number defining how a particle will respond to an electric or magnetic field that can be calculated by dividing the mass of a particle by its charge.

Massometer: In medical nanorobotics, a nanosensor device for measuring the mass of individual molecules or small physical objects to single-proton resolution.

Materials science: Science of ceramics, glass, metals, plastics, semiconductors.

MatML Materials Markup Language: Materials property data distributed on the World Wide Web in documents using hypertext markup language.

Matter as Software: "Autonomous, motile microdevices clearly are on the horizon. They may be regarded as the first step in the evolution of a technology for "programming" the structure and properties of material objects at the microscopic and the submicroscopic levels. As this evolution progresses, *the physical and economic properties of such programmable matter are likely to become much like those of present day software."* [MITRE Corporation]

MCM: MultiChip Module; the interconnection of two or more semiconductor chips in a semiconductor-type package.

Mean: Numerical average of data values.

Measurand: A physical quantity, condition, or property that is to be measured.

Meat Machine: AKA Cabinet Beast. A box containing assemblers and raw material, within which is formed meat [or whatever else it was programmed to make].

Mechanical: Pertaining to the positions and motions of atoms, as defined by the positions of their nuclei; see electronic. A purely mechanical device can be described in terms of atomic positions and motions without reference to electronic properties, save through their effect on the potential energy function.

Mechanochemistry: Chemistry accomplished by mechanical systems directly controlling the reactant molecules; the formation or breaking of chemical bonds under direct mechanical control. [See How does 'mechanochemistry' work?] In this volume, the chemistry of processes in which mechanical systems operating with atomic-scale precision either guide, drive, or are driven by chemical transformations. In general usage, the chemistry of processes in which energy is converted from mechanical to chemical form, or vice versa.

Mechanosynthesis: Chemical synthesis controlled by mechanical systems operating with atomic-scale precision, enabling direct positional selection of reaction sites; synthetic applications of mechanochemistry. Suitable mechanical systems include AFM mechanisms, molecular manipulators, and molecular mill systems. Processes that fall outside the intended scope of this definition include reactions guided by the incorporation of reactive moieties into a shared covalent framework (i.e., conventional intramolecular reactions), or by the binding of reagents to enzymes or enzymelike catalysts. Molecular tools with chemically specific tip structures can be used, sequentially, to modify a work piece and build a wide range of molecular structures. See Technical Bibliography for Research on Positional Mechanosynthesis

Mechatronics: The study of the melding of AI and electromechanical machines to make machines that are greater than the sum of their parts. The synergistic combination of precision mechanical engineering with electronic control.

Megabyte (MB): 2^{20} (= 1,048,576, or about one million) bytes of information.

Megahertz (Mhz): One million cycles per second (see also"frequency").

Meme: An idea that replicates through a society as it is propagated through person-to-person interaction, both direct and indirect. Memetics is a field of study that focuses on memes' role in the evolution of a culture. Self-reproducing idea or other information pattern which is propagated in ways similar to that of a gene. [Richard Dawkins, 1976]

MEMS MicroElectro Mechanical Systems: MEMS is less an application area in itself than a manufacturing or fabrication technique that enables other application areas. Many authors use MEMS as shorthand to imply a number of particular application areas. As it is used here, MEMS is a "top- down" fabrication technology that is especially useful for integrating mechanical and electrical systems together on the same chip. It is grouped in the category of integrated microsystems because these same MEMS techniques can be extended in the future to also help integrate biological and chemical components on the same chip, as discussed below. Thus far, MEMS techniques have been used to make some functional commercial devices such as sensors and single- chip measurement devices. Many researchers have used MEMS technologies as analytical tools in other areas of nanotechnology [Central Intelligence Agency, US The Global Technology Revolution, Chapter Two Technology Trends, Genomics, 2001]

Stood originally for Micro-ElectroMechanical System — microscopic mechanical elements, fabricated on silicon chips by techniques similar to those used in integrated circuit manufacture, for use as sensors, actuators, and other devices. Today almost any miniaturized device (based on Si technology or traditional precision engineering, chemical or mechanical) is referred to as a MEMS device generic term to describe micron scale electrical/ mechanical devices. [ZY] See The beauty of MEMS: Simpler, more reliable, cheaper, and cool - Small Times for a great description and examples of use.

MEMS: Stood originally for Micro-ElectroMechanical System — microscopic mechanical elements, fabricated on silicon chips by techniques similar to those used in integrated circuit manufacture, for use as sensors, actuators, and other devices. Today almost any miniaturized device (based on Si technology or traditional precision engineering, chemical or mechanical) is referred to as a MEMS device.

Mesoscale: A device or structure larger than the nanoscale (10^-9 m) and smaller than the megascale; the exact size depends heavily on the context and usually ranges between very large nanodevices (10^-7 m) and the human scale (1 m). [AS]

Messenger molecule: A chemically recognizable molecule which can convey information after it is received and decoded by an appropriate chemical sensor.

Metal nanoshells: A new type of *nanoparticle* composed of a **semiconductor** or dielectric core coated with an ultrathin conductive layer.. By adjusting the relative core and shell thicknesses, metal nanoshells can be fabricated that will absorb or scatter light at any wavelength across the entire visible and infrared range of the electromagnetic spectrum.

Metamorphic: In medical nanorobotics, capable of adopting multiple physical configurations via smooth changes from one configuration to another.

Metastable: A classical system is metastable if it is above its minimum-energy state, but requires an energy input before it can reach a lower-energy state; accordingly, a metastable system can act like a stable system, provided that energy inputs (e.g., thermal fluctuations) remain below some threshold. Systems with strong metastability are commonly described as stable. Quantum mechanical effects can permit metastable states to reach lower energies by tunneling, without an energy input; an associated, broader definition of *metastable* embraces all systems that have a long lifetime (by some standard) in a state above the minimum-energy state.

Micellar nanocontainers: Block copolymer micelles are water- soluble biocompatible nanocontainers with great potential for delivering hydrophobic drugs. An understanding of their cellular distribution is essential to achieving selective delivery of drugs at the subcellular level. R. Savic, L. Luo, A. Eisenberg, D. Maysinger, Micellar nanocontainers distribute to defined cytoplasmic organelles Science 300 (5619): 615- 618, Apr. 25, 2003

Micelles: Micelles are small, spherical structures composed of molecules that attract one another to reduce surface tension. The head of the molecule is hydrophilic, meaning it likes water, while the interior portion is hydrophobic, meaning it avoids water.

Microbiotagraphics: Mapping the microbiotic populations present in the human body.

Microbubbles: Very small encapsulated gas bubbles (diameters of micrometers) that can be used in diagnostic and therapeutic applications. Upon exposure to sufficiently intense ultrasound, microbubbles will cavitate, rupture, disappear, release gas content, etc. Such characteristics of the microbubbles can be used to enhance diagnostic tests, dissolve blood clots, and deliver drugs or genes for therapy. MeSH 2004

Microchemistry: The development and use of techniques and equipment to study or perform chemical reactions, with small quantities of materials, frequently less than a milligram or a milliliter. MeSH 2003

Microdevices: Narrower terms: integrated microdevices, nanodevices, SED Single Electron Devices. Related terms microelectronics, microfluidics, micro- TAS.

Micro-Electromechanical Systems (MEMS): "The fabrication or micro-machining of materials to make stationary and moving structures, devices and systems of a nominal size scale from a few centimeters to a few micrometers." *

Microelectronics: Narrower terms: MEMS,

nanoelectronics, optoelectronics, SED Single Electron Devices. Related terms: molecular electronics, semiconductors

Microencapsulation: Individually encapsulated small particles. see Journal of Microencapsulation

Microfabrication: This technology includes techniques used to manufacture integrated circuits (ICs), discrete microelectronic devices, MEMS devices such as sensors and actuators, and various electro-optic devices. [University of Louisville Microfabrication Lab] is currently serving as a center for research activity in the areas of micromachined sensors and actuators, electro- optic devices, special- purpose microelectronic devices, planar waveguides, chemical transducers, microstrip and microgap radiation detectors, micromachined nozzles, and micromachined ink- jet printheads.

Microfabrication holds the key to developing monolithically integrated devices capable of extremely high throughput. ... Microfabrication of surface characteristics provides a means of controlling the cell- environment interaction. Microfabricated structures also present new opportunities for cell, protein and nucleic acid separations and analysis- including high throughput single protein molecule sequencing. New approaches to analysis utilizing conductance techniques or tunneling probe microscopy merely emphasize the breadth of the potential of nanofabrication for future technology development relevant to biological issues. [National Center for Research Resources "Integrated Genomics Technologies Workshop Report" Jan 1999]

Microfluidics: A multidisciplinary field comprising physics, chemistry, engineering and biotechnology that studies the behavior of fluids at volumes thousands of times smaller than a common droplet. Microfluidic components form the basis of so-called "lab-on-a-chip" devices that can process microliter and nanoliter volumes and conduct highly sensitive analytical measurements. The fabrications techniques used to construct microfluidic devices are relatively inexpensive and are amenable both to highly elaborate, multiplexed devices and also to mass production. In a manner similar to that for microelectronics, microfluidic technologies enable the fabrication of highly integrated devices for performing several different functions on the same substrate chip. Microfluidics is a critical component in gene chip and protein chip development efforts. The study of fluid channels and chambers of tiny dimensions of tens to hundreds of micrometers and volumes of nanoliters or picoliters. This is of interest in biological MICROCIRCULATION and used in MICROCHEMISTRY and INVESTIGATIVE TECHNIQUES [MeSH 2004]

Enables the fabrication of networks of channels, chambers, and valves for the flow of liquids as minute as one picoliter. These systems have no moving parts and require little assembly. [Coventor "About Coventor" 2002]

Within the microelectromechanical systems (MEMS) and biological and chemical detection communities, *microfluidics* refers to the research and development of microscale devices that handle very small volumes of fluids (nano- and picoliter volumes). These devices may perform tasks such as DNA analysis or the separation of human blood cells. [Josh Molho "What is microfluidics?" Stanford Univ., US, 2000]

Microinjection: The insertion of a substance into a cell through a microelectrode. Typical applications include the injection of drugs, histochemical markers (such as horseradish peroxidase or lucifer yellow) and RNA or DNA in molecular biological studies. To extrude the substances through the very fine electrode tips, either hydrostatic pressure (pressure injection) or electric currents (ionophoresis) is employed. [OMD]

A technique for introducing a solution of DNA, protein, or other soluble material into a cell using a fine microcapillary pipet. [Life Sciences Dictionary]

Micromachines: Mechanical objects that are fabricated in the same general manner as integrated circuits. They are generally considered to be between 100 nanometres to 100 micrometres in size, though that is debatable. The applications of micromachines include accelerometers that detect when a car has hit an object and trigger an airbag. Complex systems of gears and levers are another application. The fabrications of these devices is usually done by one or both of two techniques: surface micromachining and bulk micromachining.

Micromachining: Techniques used to fabricate **MEMS**. **Broader term: microengineering**

Micron: 10^{-6} Symbol is υ. One millionth of a meter, or about 1/25,000 of an inch.

Microparticles: Applications include calibration of *flow cytometers*, particle and hematology analyzers, *confocal laser scanning microscopes* and zetapotential measuring instruments; flow measurements in gases and liquids like Laser Doppler Anemometry (LDA); Particle Dynamics Analysis (PDA), and Particle

Image Velocimetry (PIV); medical diagnostics; separation phases for *chromatography*; support for immobilized enzymes; spacer in liquid crystal displays (LCD's); peptide synthesis; cell separation; *tracers* in environmental science; model systems in medicine, biochemistry, colloid chemistry, and aerosol research. [Microparticles GmbH, Berlin Germany]

Microphone: Device that produces voltage or current in response to a sound wave.

Microprocessor: Chip containing the logical elements for performing calculations and carrying out stored instructions.

Microspectrophotometry: Analytical technique for studying substances present at enzyme concentrations in single cells, in situ, by measuring light absorption. Light from a tungsten strip lamp or xenon arc dispersed by a grating monochromator illuminates the optical system of a microscope. The absorbance of light is measured (in nanometers) by comparing the difference between the image of the sample and a reference image. MeSH 1990

Microspheres: Microspheres are generally defined as small spheres made of any material and sized from about 0.5 μm to 100 μm. Similar, but smaller spheres sized 10 to 500 nm are called *nanospheres*. Ideally, microspheres are completely spherical and homogeneous in size, although less perfect particles are often termed microspheres as well. Depending on the preparation method and material used, microspheres show a typical size distribution which often deviates from the mono-sized ideal.

Microstructure: In material engineering the structural features of a material such as grain boundaries, grain size and structure, subject to observation under a microscope, selective etching etc. In MEMS microstructure unfortunately is also used to designate a micromachined feature. The last decade has seen rapid developments in the fabrication, characterization and conceptual understanding of synthetic microstructures in many different material systems including silicon, III-V and II-VI semiconductors, metals, ceramics and organics. The objective of this journal [Superlattices and Microstructures] is to provide a common interdisciplinary platform for the publication of the latest research results on all such "nanostructures" with dimensions in the range of 1 - 100 nm; the unifying theme here being the dimensions of these artificial structures rather than the material system in which they are fabricated.

Microsystem: A microscale machine that can sense information from the environment and act accordingly. Outside the U.S., it can also refer to microelectromechanical systems (MEMS). [smalltimes glossary, 2002]

MicroTas, micro Total Analysis Systems, uTAS: Although initial research dates back to the early 1970's, the field of micro- TAS formally started in 1990, when Manz et al described the possibility of creating microsystems that would take care of many or all the traditional analytical steps involved in a biochemical analysis (sample introduction, handling, extraction, purification, concentration, filtration, analysis, detection) .... Micro- TAS offer many advantages over traditional analysis systems. Low power consumption and small reaction volumes, faster analysis, ultrasensitive detection, and minimal human intervention are key parameters in the development of micro- TAS. Most biochemical reactions take place in liquid environments. Hence, the development of MicroTAS is intrinsically linked to the design of liquid handling micro- devices. [Biomedical Applications Group (GAB) Centro Nacional de Microelectronica (CNM- IMB) Bellaterra, Spain, 2000]

Microtubule: A hollow cylindrical tube used for the transport of materials inside cells.

Miller indices: A set of 3 integers (4 for hexagonal) that designate crystallographic planes, as determined from reciprocals of fractional axial intercepts.

Millimeter: One thousandth of a meter, or about 1/26 of an inch.

MIMIC: [Micromoulding in capillaries] one-step rapid prototyping technique.

Miniaturization: Desirable for many technologies for overall cost reduction (including reduction in the amount of reagents and analytes). Important to remember that building space is often the least available and most expensive component of an overall laboratory budget.

MIPS: Millions of Instructions per Second — a measure of computing power.

Misreaction: A chemical reaction that fails by yielding an unwanted product.

MM2/CSC: A molecular mechanics program developed by Cambridge Scientific Computing that closely follows the MM2 model, adding a graphical user interface and other features.

MM2: A molecular mechanics progr am developed by Norman

Allinger and coworkers; the "MM2 model" is the molecular potential energy function described by the equations, rules, and parameters embodied in that program.

MNT: An abbreviation for molecular nanotechnology; refers to the concept of building complicated machines out of precisely designed molecules.

Mobility (electron, and hole): The proportionality constant between the carrier drift velocity and applied electric field.

Modulus of elasticity (E) : The ratio of stress to strain when deformation is totally elastic. Also the Young's modulus.

Modulus: Any of several measures of strain versus applied stress. See shear modulus, Young's modulus.

MOEMS MicroOpticalElectroMechanical systems: In addition to mechanical and electrical components, integrate waveguides or other optical features into the body of the silicon chip. Intellisense Corp. MEMS The Next Small Thing Google = about 10 Aug. 8, 2002; about 12 June 23, 2004; about 10 May 2, 2005

Moiety: A portion of a molecular structure having some property of interest.

Molality: The molality or molal concentration (symbol m) is the amount of substance per unit mass of solvent or mol kg^{-1}. Concentration in a liquid solution (symbol c), in terms of the number of moles of a solute dissolved in 10^6 mm^3 (10^3 cm^3) of solution in mol l^{-1}.

Molding (plastics): Shaping a plastic material by forcing it, under pressure at a high temperature, into a mold cavity.

Mole: A number of instances of something (typically a molecular species) equaling ~6.022 1023. *Mole* ordinarily means gram-mole; a kilogram-mole is ~6.022 1026.

Molecular - Nanotechnology: The technology of precisely-constructed molecular-scale machines; from nanometer: a billionth of a meter. For a good introduction, see What is Nanotechnology? or Ralph C. Merkle's nanotechnology page. See also my nanotechnology page. [K. Eric Drexler, *Engines of Creation*, 1986]

Molecular assembler: A general-purpose device for molecular manufacturing, able to guide chemical reactions by positioning individual molecules to atomic accuracy (e.g. mechanosynthesis) and to construct a wide range of useful and stable molecular structures according to precise specifications. Also known as an assembler, a molecular assembler is a molecular machine that can build a molecular structure from its component building blocks. [ZY]

Molecular Beam Epitaxy: [MBE] Process used to make compound (multi-layer) semiconductors. Consists of depositing alternating layers of materials, layer by layer, one type after another (such as the semiconductors gallium arsenide and aluminum gallium arsenide).

Molecular Biology: [AKA: wet nano]

Molecular Electronics (ME) [Moletronics] Any system with atomically precise electronic devices of nanometer dimensions, especially if made of discrete molecular parts rather than the continuous materials found in today's semiconductor devices. Also: Using molecule-based materials for electronics, sensing, and optoelectronics .... ME is the set of electronic behaviors in molecule-containing structures that are dependent upon the characteristic molecular organization of space .... ME behavior is fixed at the scale of the individual molecule, which is effectively the nanoscale. Any system with atomically precise electronic devices of nanometer dimensions, especially if made of discrete molecular parts rather than the continuous materials found in today's semiconductor devices. Molecular electronics offers the tantalizing prospect of eventually building circuits with critical dimensions of a few nanometers. Some basic devices utilizing molecules have been demonstrated, including tunnel junctions with negative differential resistance, rectifiers and electrically configurable switches that have been used in simple electronic memory and logic circuits. A major challenge that remains is to show that such devices can be fabricated economically using a process that will scale to circuits with large numbers of elements while maintaining their desired electronic properties. Yong Chen et. al, Nanoscale molecular-switch devices fabricated by imprint lithography, Applied Physics Letters, (82: 10): 1610- 1612 March 10, 2003 , Hewlett Packard Labs Research

Seeks to use individual molecules to perform functions in electronic circuitry now performed by semiconductor devices. Individual molecules are hundreds of times smaller than the smallest features conceivably attainable by semiconductor technology. Because it is the area taken up by each electronic element that matters, electronic devices constructed from molecules will be hundreds of times smaller than their semiconductor-based counterparts. Moreover, individual molecules are easily made exactly the same by the billions and trillions. The dramatic reduction in size, and the sheer enormity of numbers in manufacture, are the principle benefits promised by the field of molecular electronics. Also this term is most often used to describe the transport of electricity through a structure depends on the molecular energy states and the arrangement of the molecules in that structure. The structure does not necessarily have to perform a useful function.

Molecular Integrated Microsystems (MIMS): microsystems in which functions found in biological and nanoscale systems are combined with manufacturable materials. See Molecular Integrated Microsystems

Molecular machine: A mechanical device that performs a useful function using components of nanometer scale and defined molecular structure; includes both artificial nanomachines and naturally occurring devices found in biological systems. Any machine with atomically precise parts of nanometer dimensions; can be used to describe molecular devices found in nature.

Molecular Manipulator: A device combining a proximal probe mechanism for atomically precise positioning with a molecule binding site on the tip; can serve as the basis for building complex structures by positional synthesis. A programmable device able to position molecular tools with high precision, for example, to direct a sequence of mechanosynthetic steps; a molecular assembler.

Molecular manufacturing: Manufacturing using molecular machinery, giving molecule-by-molecule control of products via positional chemical synthesis, to produce complex molecular structures manufactured to precise specifications. (2) Manufacturing using molecular machinery, giving molecule-by-molecule control of products and by-products via positional chemical synthesis. Exponential general-purpose molecular manufacturing — that's a mouthful, but what does it mean? Let's take the phrase apart to see why it is so important. The building of complex structures by mechanochemical processes. The production of complex structures via nonbiological mechanosynthesis (and subsequent assembly operations).

Molecular mechanics models: Many of the properties of molecular systems are determined by the molecular potential energy function. Molecular mechanics models approximate this function as a sum of 2-atom, 3-atom, and 4-atom terms, each determined by the geometries and bonds of the component atoms. The 2-atom and 3-atom terms describing bonded interactions roughly correspond to linear springs.

Molecular medicine: A variety of pharmaceutical techniques and gene therapies that address specific molecular diseases or molecular defects in biological systems. A variety of pharmaceutical techniques and therapies in use today. Studying molecules as they relate to health and disease, and manipulating those molecules to improve the diagnosis, prevention, and treatment of disease.

Molecular mill: A mechanochemical processing system characterized by limited motions and repetitive operations without programmable flexibility (see molecular manipulator).

Molecular motors: Protein based machines that are involved in or cause movement such as the rotary devices (flagellar motor and the F1 ATPase) or the devices whose movement is directed along cytoskeletal filaments (myosin, kinesin and dynein motor families). [MeSH, 1999]

Molecular Nanogenerator: *See* Molecular Nanogenerator Developed That Can Target Cancer Cells and Destroy Then.

Molecular Nanoscience: An emerging interdisciplinary field that combines the study of molecular/ biomolecular systems with the science and technology of nanoscale structures and systems. The potential applications for this research are very broad and include such possibilities as 1) the use of biomolecules and cellular systems to self- assemble nanoelectronic circuitry and other nanoscale structures and 2) the use of lamellar host frameworks containing nano pores that can be tailored to include guest molecules for separation of chemicals for pharmaceutical and other applications. Molecular Nanoscience Alliance for Interdisciplinary Studies and Activities, Univ. of Minnesota, US,

Molecular Nanotechnology (MNT): A term coined by Drexler refering to technology resulting from the ability to thoroughly and inexpensively control matter on the molecular level. The ability to construct shapes, devices, and machines with atomic precision, and to combine them into a wide range of products inexpensively. [See Introduction to Nanotechnology] Thorough, inexpensive control of the structure of matter based on molecule-by-molecule control of products and byproducts; the products and processes of molecular manufacturing, including molecular machinery. Thorough, inexpensive control of the structure of matter based on molecule-by-molecule control of products and byproducts; the products and processes of molecular manufacturing, including molecular machinery; a technology based on the ability to build structures to complex, atomic specifications by mechanosynthesis or other means; most broadly, the engineering of all complex mechanical systems constructed from the molecular level. *See* also nanotechnology.

Molecular Recognition: A chemical term referring to processes in which molecules adhere in a highly specific way, forming a larger structure; an enabling technology for nanotechnology.

Molecular robotics: Ralph Merkle, A New Family of Six Degree of Freedom Positional Devices, Zyvex.

Molecular self-assembly: The spontaneous formation of molecules into covalently bonded, well- defined, stable structures - is a very important concept in biological systems and has increasingly become a focus of non- biological research.

Molecular sorting rotor: A class of nanomechanical device capable of selectively binding (or releasing) molecules from (or to) solution, and of transporting these bound molecules against significant concentration gradients.

Molecular surgery (molecular repair): In medical nanorobotics, the analysis and physical correction of molecular structures in the body using medical nanomachines. Analysis and physical correction of molecular structures in the body using medical nanomachines.

Molecular Systems Engineering: Design, analysis, and construction of systems of molecular parts working together to carry out a useful purpose.

Molecular Wire: A molecular wire - the simplest electronic component - is a quasi-one-dimensional molecule that can transport charge carriers (electrons or holes) between its ends. [Michael D Ward]

Molecularly imprinted polymers MIPs: A new class of materials that have artificially created receptor structures. Since their discovery in 1972, MIPs have attracted considerable interest from scientists and 3engineers involved with the development of chromatographic absorbents, membranes, sensors and enzyme and receptor mimics.

Molecule: A set of atoms linked by covalent bonds. A macroscopic piece of diamond is technically a single molecule. (Sets of atoms linked by bonds of other kinds are sometimes also termed molecules.) Group of atoms held together by chemical bonds; the typical unit manipulated by nanotechnology. The smallest particle of a chemical substance; typically a group of atoms held together in a particular patter, by chemical bonds. Molecules are groups of atoms which are bonded together. How molecules are formed is governed by the laws of physics. An illustration representing a molecule: above a gold surface. The grey balls are: carbon atoms, the little yellow balls are sulfur,: the big yellow balls are the gold,: the white are hydrogen, and the rods joining: the balls together represent the bonds.

Moledular Technology: Moledular technology promises to create a new class of imaging agents that offer distinct advantages and can be used for the early detection and diagnosis of disease. Diagnostic molecules have the potential to act as biomarkers in drug development and diagnostics, and can be used in the imaging of cancer in living subjects.

This meeting will address the challenges in implementing nanotechnology for drug delivery systems and imaging agents, and promote dialogue between diagnostic and therapeutic development. The production and application of structures, devices and systems by controlling shape and size at nanometre scale. The development and use of techniques to study physical phenomena and construct structures in the nanoscale size range or smaller.

The creation of functional materials, devices and systems through control of matter at the scale of 1 to 100 nanometers, and the exploitation of novel properties and phenomena at the same scale. Nanotechnology is emerging as a field critical for enabling essential breakthroughs that may have tremendous potential for affecting biomedicine. Moreover, nanotechnologies developed in the next several years may well form the foundation of significant commercial platforms. Emerging as a new field enabling the creation and application of materials, devices, and systems at atomic and molecular levels and the exploitation of novel properties that emerge at the nanometer scale.

Many areas of biomedicine are expected to benefit from nanotechnology including sensors for use in the laboratory, the clinic, and within the human body; new formulations and routes for drug delivery; and biocompatible, high- performance materials for use in implants.

Examples of potential uses of nanotechnology in biomedicine include the early detection and treatment of disease and the development of "smart", rejection- resistant implants that will respond appropriately as the body's needs change. Although research in this field dates back to Richard P. Feynman's classic talk in 1959, the term nanotechnology was first coined by K. Eric Drexler in 1986 in the book *Engines of Creation*. In the popular press, the term nanotechnology is sometimes used to refer to any sub- micron process, including lithography. Because of this, many scientists are beginning to use the term *molecular nanotechnology* when talking about true nanotechnology at the molecular level.

Moledulcar manufacturing: The automated building of products from the bottom up, molecule by molecule, with atomic precision. This will make products that are extremely lightweight, flexible, durable, and potentially very 'smart'.

MOLMAC: Molecular machine [Kilian, Gryphon]

Monkeywrenching: In medical nanorobotics, the mechanical or chemical jamming of cellular equilibrium processes, with a cytocidal objective. See also disequilibration.

Monoclonal antibodies: Produced by injecting animals to elicit a response from lymphocytes to produce antibodies. Lymphocytes which produce antibodies with strong binding capability can be isolated and used to produce only one kind of antibody (monoclonal) on a permanent basis once the lymphocytes are immortalized. This is accomplished by fusing them (combining them genetically) with cancer cells which have the distinction of living indefinitely in a culture. Monoclonal antibodies can be produced repeatedly and collected for use in immunodetection.

Monomer: The units from which a polymer is constructed.

Monomolecular Computing: The implantation inside a single molecule of ALL the functional groups or circuits to realize a calculation, without any help from external artifices such as re-configuration, calculation sharing between the user and the machine, or selection of the operational devices. [C. Joachim]

Moore's law: After Gordon Moore:"The number of transistors per computer chip will double roughly every two years". (2) An emperical trend in the microelectronics industry for the number of circuits per chip to double roughly every 18 months. (3) Coined in 1965 by Gordon Moore, future chairman and chief executive of Intel, it stated at the time that the of number transistors packed into an integrated circuit had doubled every year since the technology's inception four years earlier. In 1975 he revised this to every two years, and most people quote 18 months. The trend cannot continue indefinitely with current lithographic techniques, and a limit is seen in ten to fifteen years. However, the baton could be passed to nanoelectronics, to continue the trend (though the smoothness of the curve will very likely be disrupted if a completely new technology is introduced). The observation by Intel executive Gordon Moore that the number of transistors on a computer chip doubles every 1.5 years. He also observed that the number of instructions per second performed by a chip also doubles every 1.5 to 2 years. The computing power of microprocessors has been growing exponentially for the last 40 years but this cannot continue indefinitely.

MOSFET: Metal-Oxide-Semiconductor Field-Effect Transistor — device where gate electrode potential controls current flow

Multiple Cell Repair Units Working Together: Several cell repair units are shown simultaneously engaged in repairing a single neuronal cell. Communications fibers and cables link the repair units to a master controller system that directs all the repair activities from outside the scene.

Mutation: An inheritable modification in a genetic molecule, such as DNA. Mutations may be good, bad, or neutral in their effects on an organism; competition weeds out the bad, leaving the good and the neutral.

MUX: Device for combining several signals or data streams into a single flow.

Myosin: Myosin is one of the two major proteins responsible for contraction of muscle. In muscle cells, myosin is arranged in long filaments that lie parallel to the microfilaments of actin. In muscle contraction, filaments of actin and myosin alternately unlink and chemically link in a sliding action. The energy for this reaction is supplied by adenosine triphosphate.

N

N terminus: *See* amino terminus

Na+-K+ ATPase: *See* sodium pump

NAD+ (nicotine adenine dinucleotide)**:** Coenzyme that participates in an oxidation reaction by accepting a hydride ion (H-) from a donor molecule. The NADH formed is an important carrier of electrons for oxidative phosphorylation. Coenzyme closely related to NAD+ that is used extensively in biosynthetic, rather than catabolic, pathways.

Nanapheresis: In medical nanorobotics, the removal of bloodborne medical nanorobots from the body using aphersis-like processes.

Nanarchist: Someone who circumvents government control to use nanotechnology, or someone who advocates this. [Eli Brandt, October 1991]

Nanarchy: The use of automatic law-enforcement by nanomachines or robots, without any human control - see blue goo [Mark S. Miller].

NAND: NOT-AND — logic gate whose output is the negation of that of the AND gate.

Nanite: Machines with atomic-scale components.

(Popularized by the Star Trek episode "Evolution") As to their weight, a popular question: "Do you 'feel' heavier after you drink a mouthful of water? A mouthful of water, roughly 5 cm^3, would have the same mass as a ~2 terabot (2 trillion nanites) dose of 1 micron^3 nanorobots. You'll never feel it." Robert A. Freitas Jr. "Nanobot" and "Nanorobot" usually mean the same thing. Slang term for a nanomachine (see nanotechnology), esp. a machine able to replicate itself. [Popularized by the Star Trek episode "Evolution"]

Nanny: A cell-repair nanite

Nano- : A prefix meaning one billionth (1/1,000,000,000).

Nano Cubic Technology: An ultra-thin layer coating that results in higher resolution for recording digital data, ultra-low noise and high signal-to-noise ratios that are ideal for magneto-resistive (MR) heads. It is capable of catapulting data cartridge and digital videotape to one-terabyte native (uncompressed) capacities and floppy disk capacities to three gigabytes. To help visualize the potential, 1TB can store up to 200 two-hour movies.

Nanoarray: an ultra-sensitve, ultra-miniaturized array for biomolecular analysis. BioForce Nanosciences' Nanoarrays utilize approximately 1/10,000th of the surface area occupied by a conventional microarray, and over 1,500 nanoarray spots can be placed in the area occupied by a single microarray domain. [BioForce Nanosciences] See The NanoPro™ System

Nanoassembler: the Holy Grail of nanotechnology; once a perfected nanoassembler is availble, building anything becomes possible, with physics and the imagination the only limitation (of course each item would have to be designed first, which is another small hurdle). See Bootstrapping a Nanofactory "So how hard is it to build a nanofactory? You need to start with a working fabricator, a nanoscale device that can combine individual molecules into useful shapes. But once you have that, the rest is pretty straightforward." —CRN

Nanobalance: Simply put, a nanoscale balance for determining mass, small enough to weigh viruses and other sub-micron scale particles. *"A mass attached at the end of a nanotube shifts its resonance frequency. If the nanotube is calibrated (i.e., its spring constant known), it is possible to measure the mass of the attached particle."* A nanobalance *"could be useful for determining the mass of other objects on the femtogram to picogram size range."* See Weighing The Very Small.
Transmission electron microscope image shows "nanobalance:" a resonating carbon nanotube used to determine the mass of the attached carbon particle.

Nanobarcode: SurroMed's Nanobarcode™ technology uses cylindrically-shaped colloidal metal nanoparticles, in which the metal composition can be alternated along the length and the size of each metal segment can be controlled. Intrinsic differences in reflectivity between the metal segments allow individual particles to be identified by conventional optical microscopy. See SurroMed and Penn State Publish Pioneering Work in the Journal Science Describing Nanobarcode™ Particles

Nanobeads: Polymer beads with diameters of between 0.1 to 10 micrometers. Also called nanodots, nanocrystals and quantum beads. *Impregnating fluorescent crystal chips into these beads allows*

simultaneous measurement of thousands of biological interactions, a stepping stone for breakthroughs in the diagnosis and treatment of disease. ... with the potential to accelerate drug discovery and clinical diagnostics." See Nanodots and Local Mechanical Properties of Cells.

Nanobiology: Many fundamental biological functions are carried out by molecular machineries that have the sizes of 1-100 nm. You find many examples in molecular biology and cell biology: single enzymes, transcription complex, ribosome, transport complex, nuclear pore, and so on. To understand the functions of these machineries, one has to describe their movements, changes in their shapes, and their localization. This means the mechanistic study is equivalent to dynamic morphology at this level of size, making a new field that merges mechanistic biology and morphology. The emergence of nanobiology depended on the invention of nanotechnology: scanning probe microscopy, modern optical techniques, and micro- manipulating techniques. This concept of nanobiology was first proposed in a group study named "Biological Nano-Mechanisms", which was supported by Japanese Agency of Science and Technology (1992-1997). [National Institute of Genetics, Japan]

Nanobioprocessors: Implantable nano scale processors that can integrate with biological pathways and modify biological processes. Trans- NIH Bioengineering Nanotechnology Initiative, SBIR, PA Number 02- 125:

Nanobiotechnology: An emerging area of scientific and technological opportunity. Nanobiotechnology applies the tools and processes of nano/ microfabrication to build devices for studying biosystems. Researchers also learn from biology how to create better micro-nanoscale devices. The Nanobiotechnology Center (NBTC), a National Science Foundation, Science and Technology Center is characterized by its highly interdisciplinary nature and features a close collaboration between life scientists, physical scientists, and engineers. [NBTC, NanoBiotechnology Center, Cornell Univ. US, 2002]

Use of nanotechnology[ies] in the life sciences. These may include, but are not limited to, therapeutics, medical devices/implants, biosensors, and tools for the development of drugs. Applying the tools and processes of MNT to build devices for studying biosystems, in order to learn from biology how to create better nanoscale devices. Should hasten the creation of useful micro devices that mimic living biological systems.

Nanobot: Machines with atomic-scale components. (Popularized by the Star Trek episode "Evolution") As to their weight, a popular question: "Do you 'feel' heavier after you drink a mouthful of water? A mouthful of water, roughly 5 cm^3, would have the same mass as a ~2 terabot (2 trillion nanites) dose of 1 micron^3 nanorobots. You'll never feel it." Robert A. Freitas Jr. "Nanobot" and "Nanorobot" usually mean the same thing. Slang term for a nanomachine (see nanotechnology), esp. a machine able to replicate itself. [Popularized by the Star Trek episode "Evolution"]

Nanobubbles: tiny air bubbles on colloid surfaces. Thought to reduce drag, such as would be of benefit to swimmers wearing a suit coverd in them. See First Images of Nanobubbles

Nanocantilever: The simplest micro-electro-mechanical system (MEMS) that can be easily machined and mass-produced via the same techniques used to make computer chips. The ability to detect extremely small displacements make nanocantilever beams an ideal device for detecting extremely small forces, stresses and masses. Nanocantilevers coated with antibodies, for example, will bend from the mass added when substrate binds to its antibody, providing a detector capable of sensing the presence of single molecules of clinical importance.

Nanocatalysis See LBL-UCB Scientists Demonstrate Nanocatalysis and Nanocluster arrays refine the catalytic process

Nanocell Surgery: Modifying cellular structures by the use of medical nanomachines.

Nanocentrifuge: In medical nanorobotics, a proposed nanodevice that can spin materials at very high speed, imparting rotational accelerations of up to one trillion gravities (g's), thus permitting rapid sortation.

Nanocentrifuge: In medical nanorobotics, a proposed nanodevice that can spin materials at very high speed, imparting rotational accelerations of up to one trillion gravities (g's), thus permitting rapid sortation.

Nanochemistry: The goal of the Nanochemistry group is to develop chemistry beyond the realm of individual molecules by exploring how systems of several molecules or other electronically confined entities self-organize into distinct structures on the nanometer scale. The new insight is used to study and manipulate biological nanosystems and to create novel nanoarchitectures with unique structural, electronic, optical and magnetic properties. Specific research areas include bionanotechnology and self-assembling molecular electronics.

Nanochip: In Friday's issue [Aug. 1, 2001] of the journal *Science,* physicists from IBM's Thomas J. Watson Research Center [Philip G. Collins, Michael S. Arnold and Phaedon Avouris] announce their fabrication of the world's first array of transistors made from carbon nanotube. We are approaching the limits of standard microchip technology; thus, the "nanochip" — a next-smaller microchip. [ed] They are also a next-gen device for mass storage, of significantly higher density, with greater speed, and much lower cost. [Tod Maffin (p)] See Getting a Line on Nanochips

Nanochondria: Nanomachines existing inside living cells, participating in their biochemistry (like mitochondria) and/or assembling various structures. See also nanosome. [Ken Clements 1996]

Nanochronometer: In medical nanorobotics, a proposed clock or timing mechanism constructed of nanoscale components.

Nanocircles: Stanford scientists have synthesized a molecule of DNA that is capable of shutting off specific genes in living bacteria. Dubbed the "nanocircle," the new nanometer- size molecule might one day give researchers the ability to target harmful genes that cause cancer and other diseases in humans. A nanocluster or nanocrystal is a fragment of solid comprising somewhere between a few atoms to a few tens of thousands of atoms. Nanoclusters are therefore a novel state of matter with properties that are neither those of a bulk crystal nor those of individual atoms and molecules. Over the past 10 years huge advances have been made both in the synthesis of size- tunable, monodisperse (i.e. size-selected) nanoclusters of various chemical compositions and in the development of techniques for their assembly into nanostructured solids (facilitating the synthesis of what have been termed "designer materials"). Nanoparticles, Nanoclusters, Nanoscience Group, Nottingham Univ., UK, 2002

Nanocomputer: A computer made from components (mechanical, electronic, or otherwise) built at the nanometer scale. These computers could be many orders-of-magnititude faster than today's, which enables software to take proportional leaps. (2) A computer whose fundamental components measure only a few nanometers in size. State of the art current computer components are no smaller than about 350 nm. [Mnemosyne Mnews 21 (3) January 2001 " The Nanotechnology Initiative and Future Electronics" Presentation by Gail J. Brown, Air Force Research Laboratory, Wright- Patterson Air Force Base, November 16, 2000] (3) A computer with parts built on a molecular scale.

Nanocones: Nonplanar graphitic structures. Carbon-based structures with five-fold symmetry that form due to disclination defects in two-dimensional graphene sheets. They have been observed as nanotube caps and as freestanding structures. [North Carolina State University] see Nanostructures Fabrication from Carbon Nanocones

Nanocontainers: "Micellar nanocontainers" or "Micelles," these are nanoscale polymeric containers that could be used to selectively deliver hydrophobic drugs to specific sites within individual cells. See Nanocontainers deliver on drugs.

Nanocrit (Nct)**:** In medical nanorobotics, volume-fraction or bloodstream concentration of medical nanorobots, expressed as a percentage.

Nanocrystals: A nanocrystal typically has a diameter of between 1 and 10 nm and may contain as few as a hundred or as many as tens of thousands of atoms. Many fundamental properties of nanocrystals depend strongly on their size in smooth and predictable ways. Examples include the external field required to switch a magnetized particle of great importance in magnetotactic bacteria and in hard disk drives and the color of light emission from a semiconductor used for the fluorescent *labeling* of cells and in lasers. This facile tuning of properties by size variation is one reason why nanocrystals are widely viewed as promising components for new artificial optical and electrical materials.

Also known as nanoscale semiconductor crystals. "Nanocrystals are aggregates of anywhere from a few hundred to tens of thousands of atoms that combine into a crystalline form of matter known as a "cluster." Typically around ten nanometers in diameter, nanocrystals are larger than molecules but smaller than bulk solids and therefore frequently exhibit physical and chemical properties somewhere in between. Given that a nanocrystal is virtually all surface and no interior, its properties can vary considerably as the crystal grows in size." [LBL] See Nanocrystals: The shapes of things to come

The first atomic-scale images of nanocrystals that help reduce pollution show a surprising triangular, rather than hexagonal, shape. The new information should help researchers improve the chemical process.

"Nanocrystals might be used to make super-strong and long-lasting metal parts. The crystals also might be added to plastics and other metals to make new types of composite structures for everything from cars to electronics." See Discovery could bring widespread uses for 'nanocrystals'. Single atoms caged inside nanocrystals gives you a "quantum confined atom", or QCA, *"with potential uses ranging from clear-glass sunglasses to bio-sensors to optical computing and just about anything optical in between."* Single-electron transistor (SET) is a three terminal device, where single electron current between a source and a drain through a nanocrystal is controlled by a gate. The nanocrystals are the tiny light specs.

Nanodefenses: Any of the "good" goo's, such a Blue Goo. Protectors against Grey Goo, destructive nanoswarms, and the like.

Nanodevices: Scientists believe nanoscale devices may lead to computer chips with billions of transistors, instead of millions - which is the typical range in today's semiconductor technology. The more transistors crammed on a chip, the more powerful it is. "This technology has the potential to replace existing manufacturing methods for integrated circuits, which may reach their practical limits within the next decade when Moore's Law eventually hits a brick wall," said physicist Bernard Yurke of Bell Labs. DNA, which provides the molecular blueprints for all living cells, is an ideal tool for making nanoscale devices. "We took advantage of how pieces of DNA - with its billions of possible variations - lock together in only one particular way, like pieces of a jigsaw puzzle," Yurke said.

["Researchers from Lucent Technologies' Bell Labs and University of Oxford create first DNA motors" Lucent Technologies press release, Aug. 9, 2000]

Nanodisaster: see the various 'goo' scenerios that have potentially negative outcomes. NEMS - nanoelectromechanical systems: A generic term to describe nano scale electrical/mechanical devices.

Nanoelectronics: Electronics on a nanometer scale, whether made by current techniques or nanotechnology; includes both molecular electronics and nanoscale devices resembling today's semiconductor devices. [NTN]

Nanoengineering: We use chemistry to construct **nanostructures** and their composites, then focus our attention on the electronic, optical, and transport properties of these nanostructures and the macroscopic films and materials that can be constructed from them. This research lies at the common frontier of chemistry, condensed matter physics, optics, and bioengineering.

Nanofabrication: Construction of items using assemblers and stock molecules. see Nanofacture. AKA: nanoscale engineering. Nanofabrication methods can be divided into two categories: top- down methods, which carve out or add aggregates of molecules to a surface, and bottom- up methods, which assemble atoms or molecules into nanostructures. [George M. Whitesides and J. Christopher Love "The art of building small" Scientific American 285 (3): 39- 47, Sept. 2001]

Nanofactory: A self-contained macroscale manufacturing system, consisting of many molecular manufacturing systems feeding a convergent assembly system.

Nanofacture: The fabrication of goods using nanotechnology [Geoff Dale 1995].

Nanofibers, nanofibres: Using a proprietary process, eSpin is able to produce minute fibers which are 10 to 100 times smaller in diameter than what is possible with conventional textile technology. For comparison, eSpin nanofibers are about 100 times smaller than a human hair. These nanofibers provide a very large surface area for a given weight of fibers. High surface area is at the heart of many envisioned products. Used for clean room products, nanocomposites, filtration, surgical gowns, biomedical devices, and specialty fabrics. [eSpin Technologies, "Products"].

Nanofilters: One opportunity for nanoscale filters is for the separation of molecules, such as proteins or DNA, for research in genomics. See Selective nanofilters for proteins, DNA Another, as *"masks to prevent exposure to biological pathogens such as viruses that can be as small as 30 nanometers in diameter."* See Biologically inspired nanotechnology. And another use is in water filtration. See Softer, Purer Water.

Nanofluidics: Controlling nano-scale amounts of fluids Researchers at Cornell University are using nanotechnology to build microscopic silicon devices with features comparable in size to DNA, proteins and other biological molecules — to count molecules, analyze them, separate them, perhaps even work with them one at a time.

Nanogate: A device that precisely meters the flow of tiny amounts of fluid. Precise control of the flow restriction is accomplished by deflecting a highly polished cantilevered plate. The opening is adjustable on a sub-nanometer scale, limited by the roughness of the polished plates. Thus, the Nanogate is an Ultra Surface Finish Effect Mechanism (USFEM). The Nanogate can be fabricated on a macro-, meso- or micro- (MEMs) scale. [James R. White] : See Nanogate: A Fundamental New Device for Nanofluidics

Nanogenomics: The main objective of the group is to develop experimental tools that will accelerate the finding of genomic factors (e.g. genes, polymorphisms, methylation patterns) that contribute to complex traits. In the long term this may lead to predictive medicine based on personalized genomic information. Currently available molecular tools are too slow, cumbersome and expensive to allow comprehensive genomic information to be easily obtained from a large number of individuals in a population (e.g. for case/control studies). A radical re-invention of the way genes and DNA are analyzed is needed. We believe that the analysis of single DNA molecules can overcome many of the present limitations and in addition will reveal information which is inaccessible by conventional methods.

Nanoguitar: "Made for fun to illustrate the technology — the world's smallest guitar is 10 micrometers long — about the size of a single cell — with six strings each about 50 nanometers, or 100 atoms, wide. Just one of several structures that Cornell researchers believe are the world's smallest silicon mechanical devices. Researchers made these devices at the Cornell Nanofabrication Facility, bringing microelectro-mechanical devices, or MEMS, to a new, even smaller scale — the nano-sized world." See World's smallest silicon mechanical devices are made at Cornell. Smallest guitar, about the size of a human blood cell, carved out of crystalline silicon, illustrates new technology for nanosized electromechanical devices.

Nanogypsy: someone who travels form place to place, spreading the "nano" word. Usually a person who takes the most optimistic viewpoint, and is enthusitic.

Nanohacking: describes what MNT is all about — "hacking" at the molecular level. :

Nanoharvesting: Nanoharvesting agents are being designed to act as molecular mops for these [cancer] biomarkers and can be directly queried by mass spectrometry. Dr. Emanuel F. Petricoin, Co-Director, NCI-FDA Clinical Proteomics Program, Senior Principal Investigator, Center for Biologics Evaluation and Research, U. S. Food and Drug Administration.

Nanohorns: One of the SWNT (single walled carbon nanotube) types, with an irregular horn-like shape, which may be a critical component of a new generation of fuel cells. "The main characteristic of the carbon nanohorns is that when many of the nanohorns group together an aggregate (a secondary particle) of about 100 nanometers is created. The advantage being, that when used as an electrode for a fuel cell, not only is the surface area extremely large, but also, it is easy for the gas and liquid to permeate to the inside. In addition, compared with normal nanotubes, because the nanohorns are easily prepared with high purity it is expected to become a low-cost raw material." See NEC uses Carbon Nanotubes to Develop a Tiny Fuel Cell for Mobile Applications and here is a TEM image.

Nanoimprint Lithography (NIL): A method of

lithography that uses a mold, or mechanical force (embossing) to pattern a resist.

Nanoimprint Machine: a form of soft lithography.

Nanoimprinting: Sometimes called soft lithography. A technique that is very simple in concept, and totally analogous to traditional mould- or form-based printing technology, but that uses moulds (masters) with nanoscale features. As with the printing press, the potential for mass production is clear. There are two forms of nanoimprinting, one that uses pressure to make indentations in the form of the mould on a surface, the other, more akin to the printing press, that relies on the application of "ink" applied to the mould to stamp a pattern on a surface. Other techniques such as etching may then follow. Using an imprinting technique to create nanostructures. With photolithography successfully producing sub-100-nm structures, attention has now turned to 50-nm structures. UCI Division Director Yee Discusses Nanoimprinting at UCSD

Sometimes called soft lithography. A technique that is very simple in concept, and totally analogous to traditional mould- or form-based printing technology, but that uses moulds (masters) with nanoscale features. As with the printing press, the potential for mass production is clear. There are two forms of nanoimprinting, one that uses pressure to make indentations in the form of the mould on a surface, the other, more akin to the printing press, that relies on the application of "ink" applied to the mould to stamp a pattern on a surface.

Nanoindentation: Nanoindentation is similar to conventional hardness testing performed on a much smaller scale. The force required to press a sharp diamond indenter into a material is measured as a function of indentation depth. As depth resolution is on the scale of nanometers (hence the name of the instrument), it is possible to conduct indentation experiments even on thin films. Two quantities which can be readily extracted from nanoindentation experiments are the material's modulus, or stiffness, and its hardness, which can be correlated to yield strength. Investegators have also used nanoindentation to study creep, plastic flow, and fracture of materials. [Nix Research Group, Materials Science & Engineering, Stanford University]

Nanolithography: Writing on the nanoscale. From the Greek words Nanos - Dwarf, Lithos - rock, and grapho - to write, this word literally means "small writing on rocks." [NTN] see Optical Near-Field Lithography

Nanomachine: An artificial molecular machine of the sort made by molecular manufacturing. Mechanical devices so small that the parts are single molecules. One approach to manufacturing nanomachines involves using biological molecules — such as DNA, RNA, enzymes and proteins — to synthesize and duplicate useful devices (this might be termed the bottom-up approach). The other major approach involves miniaturizing today's microfabrication tools by stages, ultimately to work at the nanoscale (the top- down approach). Each approach has strengths and limitations, and the first nanofactories capable of molecular manufacturing will likely use some combination of both. David B. Hughes, Nanoassemblers to Nanofactories, Nanobiology, 2005 http://techbio.info/nanobiology/node/5

Nanomachining: like traditional machining, where portions of the structure are removed or modified, nanomachining involves changing the structure of nano-scale materials or molecules.

Nanomanipulation: The process of manipulating items at an atomic or molecular scale in order to produce precise structures. [ZY See Zyvex SEM manipulator]

NanoManipulator: uses virtual reality (VR) goggles and a force feedback probe as an interface to a scanning probe microscope, providing researchers with a new way to interact with the atomic world. Researchers can travel over genes, tickle viruses, push bacteria around, and tap on molecules - the nanoManipulator simplifies the process and allows researchers to play with their atoms. University of North Carolina at Chapel Hill (UNC-CH) *The Nanomanipulator* from the Center for Computer Integrated Systems for Microscopy and Manipulation (CISMM) at UNC Chapel Hill. Part of the Nanoscale Science Research Group (NSRG).

Nanomanufacturing: Same as molecular manufacturing. *See* Molecular manufacturing. Is expected to be high- volume, high- rate, integrated assembly of nano- elements into commercial products. This involves controlling position, orientation, and interconnectivity of the nano- elements. Increases in worldwide investments over the past few years have propelled nanoscience research scientific breakthroughs to a new level. To ensure that these discoveries lead to commercially viable products, it is important to address fundamental scientific barriers to nanomanufacturing, in parallel with the ongoing nanoscience research.

Nanomaterials: Can be subdivided into nanoparticles, nanofilms and nanocomposites. The focus of nanomaterials is a bottom up approach to structures and functional effects whereby the building blocks of materials are designed and assembled in controlled ways. [Oxonica] Contain only a few thousand or tens of thousands of atoms, rather than the millions or billions of atoms in particles of most conventional materials. *Nanostructures* can be obtained in metallic, ceramic, semi-conducting, and diamond materials. Applications, Nanopowders Industry.

This area combines nanotechnology and many applications of nanostructured materials. One important research area is the formation of semiconductor "*quantum dots*" (i.e., several nanometer- size, faceted crystals) by injecting precursor materials conventionally used for chemical-vapor deposition of *semiconductors* into a hot liquid surfactant. This "quantum dot" is in reality a macromolecule because it is coated with a monolayer of the surfactant, preventing agglomeration. These materials photoluminesce at different frequencies (colors) depending upon their size, allowing optical multiplexing in biological labeling.

Nanomechanical: Being mechanical and very small; for example, a robot that can manipulate single molecules. Pertaining to nanomachines.

Nanomedicine: (1) the comprehensive monitoring, control, construction, repair, defense, and improvement of all human biological systems, working from the molecular level, using engineered nanodevices and nanostructures;

(2) the science and technology of diagnosing, treating, and preventing disease and traumatic injury, of relieving pain, and of preserving and improving human health, using molecular tools and molecular knowledge of the human body;

(3) the employment of molecular machine systems to address medical problems, using molecular knowledge to maintain and improve human health at the molecular scale.

(4) A rapidly expanding field that includes many potential technologies and approaches. The key to this definition is that phenomena and materials at the nanometer scale are known to have properties that are uniquely attributable to that scale length. Nanomedicine could similarly be defined [as nanotechnology] as the design, synthesis, or application of materials, devices, or technologies in the nanometer-scale for the basic understanding, diagnosis, and / or treatment of disease. Canadian Institute of Health Research, Regenerative Medicine and Nanomedicine, RFA, 2003. The monitoring, repair, construction and control of human biological systems at the molecular level, using engineered nanodevices and nanostructures.

(5) the comprehensive monitoring, control, construction, repair, defense, and improvement of all human biological systems, working from the molecular level, using engineered nanodevices and nanostructures;

(6) the science and technology of diagnosing, treating, and preventing disease and traumatic injury, of relieving pain, and of preserving and improving human health, using molecular tools and molecular knowledge of the human body;

(7) the employment of molecular machine systems to address medical problems, using molecular knowledge to maintain and improve human health at the molecular scale. The use of molecular-scale devices (nanotechnology) to repair damage and boost

the immune system. Vapour grown carbon fibres are obtained as shown in this SEM image. The diameters of these fibres can vary from 100 nm to 500 nm. As part of the National Institutes of Health (NIH) Roadmap for Medical Research [nihroadmap.nih.gov], the NIH [nih.gov] will establish a handful of nanomedicine centers. These centers will be staffed by a highly interdisciplinary scientific crew including biologists, physicians, mathematicians, engineers and computer scientists. Research conducted over the first few years will be spent gathering extensive information about how molecular machines are built. A key activity during this time will be the development of a new kind of vocabulary, or lexicon, to define biological parts and processes in engineering terms.

Once researchers have completely catalogued the interactions between and within molecules, they can begin to look for patterns and a higher order of connectedness than is possible to identify with current experimental methods. Mapping these networks and understanding how they change over time will be a crucial step toward helping scientists understand nature's rules of biological design. Understanding these rules will, in many years' time, enable researchers to use this information to address biological issues in unhealthy cells.

Nanomesh and Nanofibres: (or "Nanofibers") This term covers CNT's (see above), and as described here, the other "nanoscale fibers" referred to as "polymeric" (made from polymers). Currently used in air and liquid filtration applications. Using a process called "electrospinning" - or e-spin - a polymer "mesh" is formed into a nanofiber membrane, hense "nanomesh", with 150 - 200 nm diameters. Some have been made since 1970, but were not called "nano" until recently. One potential use is *"to prevent body tissues from sticking together as they heal. It also breaks down in the body over time like biodegradable sutures."* , which makes it a surgical material for the 21st Century. Other uses include biomedical devices, filtration systems, and dust collecting systems. See Biodegradable nanofiber could prevent scar tissue.

Nanometals: Metal particles of diameter in the range of a few nanometers or thin films in the same thickness range, are interesting not only because of their special mechanic properties but also because of other physical and chemical properties, sometimes totally different from those of coarse-grained metals. Metal based magnetic materials are of great interest for the field of information storage. Encapsulated (for example in a carbon matrix) metals are protected against oxidation, without loosing their magnetic properties. Encapsulated or supported nanometals are more resistant against sintering at elevated temperatures. Metallic thin films may find their application in electronic industry, for example as interconnect lines or magnetic or electric layers. Metal Based Nanomaterials, E-MRS European Materials Research Society, Fall 2004 meeting

Have a wide variety of uses including energetics (rocket propulsion and pyrotechnics), in microelectronic films and coatings, super conducting alloys and for high strength powder metallurgical metals and alloys. Any metal that is available as a continuous ductile wire can be converted by our process into nano metal spheres. Nanomaterial Technologies, Argonide

Nanometer (nm): Unit of length commonly used to measure molecules and cell organelles. 1 nm = 1E-3 μm = 1E-9 m. A unit of spatial measurement that equals one-billionth (10 -9) of a meter. The head of a pin is about 1 million nanometers across. A human hair is about 60,000 nanometers in diameter, while a DNA molecule is between 2-12 nanometers wide. One billionth of a meter; approximately the length of three to six atoms placed side-by-side, or the width of a single strand of DNA; the thickness of a human hair is between 50,000 and 100,000 nanometers. One billionth of a metre.

Nanomotors: A University of Florida chemistry professor has made a "nanomotor" from a single DNA molecule. The motor, so small that hundreds of thousands could fit on the head of a pin, curls up and extends like an inchworm, said Weihong Tan, the principal investigator and lead author of an article about the motor in the April edition of the journal *Nano Letters*.

Nanonewtons: Forces 1 billion times smaller than the force required to hold an apple against Earth's gravity. Nanonewton forces are estimated with atomic force microscopes and instruments that measure the properties of ultrathin coatings like those used on computer hard drives or turbine blades. Nanotechnology: Cracking the nanonewton force barrier NIST, US, 2003.

Nano-Optics: Interaction of light and matter on the nanoscale. see University of Rochester - Institute of Optics - NANO-OPTICS GROUP

Nanoparticle: A nanoscale spherical or capsule-shaped structure. Most, though not all, nanoparticles are hollow, which provides a central reservoir that can be filled with anticancer drugs, detection agents, or chemicals, known as reporters, that can signal if a drug is having a therapeutic effect. The surface of a nanoparticle can also be adorned with various targeting agents, such as antibodies, drugs, imaging agents, and reporters. Most nanoparticles are constructed to be small enough to pass through blood capillaries and enter cells.

The use of commercially available nanoparticle constructs for both imaging and drug delivery in humans is not novel. However, the current enthusiasm for nanotech-based solutions extends far beyond simple particles to multi-modal platforms that involve multiple active pharmaceutical ingredients (APIs). These new constructs present a higher level of complexity not only in manufacturing but also in predicting their behavior in the human including such fundamental information as bio-distribution and biocompatibility. Thus go/ no-go decision points are less well defined and decisions such as committing resources for scale up of GLP product for pre-clinical testing must be made on limited data. "The Bio-Legal Complexity of Nanoparticle Development" Dr. James L. Tatum, Special Assistant, Cancer Imaging Program, National Cancer Institute

Nanoparticles, including *nano- clusters, [nano]-layers, [nano]- tubes,* and two- and three- dimensional structures in the size range between the dimensions of molecules and 50 nm (or in a broader sense, submicron sizes as a function of materials and targeted phenomena), are seen as tailored precursors for building up functional *nanostructures.* [R&D Status and Trends in Nanoparticles, Nanostructured Materials, and Nanodevices in the US, Proceedings of the May 8-9, 1997 Workshop, Jan. 1998 Richard W. Siegel, WTEC Panel Chair]

Nanopens & Nanopencils: (AKA: Atomic Pencil) *"Analogous to using a quill pen but on a billionth the scale"*, and may transform dip-pen nanolithography. Allows for drawing electronic circuits a thousand times smaller than current ones. The "pen" is an atomic force microscope (AFM). See Nanopipettes and Nanoplotter for further details.

NanoPGM-nanometer-scale patterned granular motion: The goal of NanoPGM is to generate millions of ìnanofingers, finger-like structures each only a few nanometers long, that might someday perform precise, massively parallel manipulation of molecules and directed assembly of other nanometer-scale objects. This ability answers one of the biggest technical challenges facing builders of nanocomputers: how to arrange as many as a trillion molecular computing components in an area only a few millimeters square. [MITRE / Alex Wissner-Gross]

Nanopharmaceuticals: Nanoscale particles used to modulate drug transport for drug uptake and delivery applications.

Nanophase Carbon Materials (carbon nanotubes, nanodiamond, nanocomposite]**:** A form of matter in which small clusters of atoms form the building blocks of a larger structure. These structures differ from those of naturally occurring crystals, in which individual atoms arrange themselves into a lattice.

Nanophotonics: The ongoing development of new techniques for fabricating nanoelectronic devices opens up the possibility of using these methods for making nanophotonic devices. That is, devices that involve the interaction of light with features only tens or hundreds of nanometers in size. We are currently doing exploratory research to investigate such effects and related devices that may find future use in information technology. One example is the application of interferometric lithography to the fabrication of photonic- crystal- like structures. The nanoscale physics group uses various experimental techniques to examine the physical properties of objects in the nanoscale size range, that is, a little bit larger than the size of atoms. Some interesting physical properties at this range include conductivity of small numbers of atoms and molecules, forces arising between objects on this scale, and the transition between the quantum nature of a few atoms and a large number of atoms. [Nanoscale Physics, Purdue Univ., 2000] http://www.physics.purdue.edu/nanophys/

Nanopipettes: "Cantilevered/Straight Nanopipettes can be used as *nanopens* for controlled chemical delivery or removal from regions as small as 100 nanometers. They can also be used as vessels for containing molecules whose optical properties change in response to their chemical environment." Other uses include "controlled chemical etching with the precision of atomic force microscopy; chemical imaging of surfaces; delivering femtosecond laser pulses; and performing NSOM/SNOM imaging using a UV excimer laser." See Cantilevered/Straight Nanopipettes Modifying the nanopipette yields other nanotools, such as Nanotweezers and Nanoheaters. See Nanotools.

Nanoplotter: A multi-tip nanopen. "A device that can draw patterns of tiny lines just 30 molecules thick and a single molecule high. ... produces eight identical patterns at once and extends ... dip-pen nanolithography towards mass producing nanoscale devices and circuits by converting what was a serial process to a parallel one. May be use to "... miniaturize electronic circuits, pattern precise arrays of organic and biomolecules such as DNA and put thousands of different medical sensors on an area much tinier than the head of a pin." See Plotting Chemicals and Nanoplotter with Parallel Writing Capabilities.

Nanoplumbing: Illustrations of projects that would be considered relevant to this PA. o Nanoplumbing components such as valves, microfluidic channels, and motors (e.g., to be used as pumps).

Nanopores: Involves squeezing a DNA sequence between two oppositely charged fluid reservoirs, separated by an extremely small channel. Essentially itty bitty tiny holes. Nanoscopic pores found in purpose-built filters, sensors, or diffraction gratings to make them function better. See Influencing structure in the heart of nanoland. As activated carbon, they may also be used as an alternative fuel storage medium, due to their massive internal surface area. "Scientists believe nanopores, tiny holes that allow DNA to pass through one strand at a time, will make DNA sequencing more efficient." See Understanding Nanodevices — Nanopores. In biology, they are "complex protein assemblies that span cell membranes and allow ionic transport across the otherwise impermeable lipid bilayer. Nanopores are important because while some pores help maintain cell homeostasis, others disrupt cell function." See Towards Fabrication of Solid-State Mimics of Biological Nanopores. "A nanopore can be a protein channel in a lipid bilayer or an extremely small isolated 'hole' in a thin, solid-state membrane" such that "DNA and RNA, can be registered and characterized singly ..."

Nanoprobe: Nanoscale machines used to diagnose, image, report on, and treat disease within the body. See "Cell Repair Machine", "Nanites", "Nanobots", and "Nanomachine". Also: tips for scanning probe microscopes.

Nanoreplicators: A set of nanomachines capable of exponential replication.

Nanorobot: A computer-controlled robotic device constructed of nanometer-scale components to molecular precision, usually microscopic in size (often abbreviated as "nanobot").

Nanorods: or Carbon Nanorods. Formed from multi-wall carbon nanotubes. Another nanoscale material with unique and promising physical properties, such that may yield improvements in high-density data storage, and allow for cheaper flexible solar cells. See Three Element Nanorods and Flexible and Inexpensive Solar Cells Based on Inorganic Nanorods.

Nanoropes: nanotubes connected and strung together.

Nanoscale: 1-100 nanometer range.

Nanoscale: On a scale of nanometers, from atomic dimensions to ~100 nm.

Nanoscale: Significantly smaller than a micron; on the scale of large molecules; capable of interacting with molecules; capable of being built by a single assembler.

Nanoscience: Nanoscience is the science of nanoscale

phenomena. This is a very complicated area, since Quantum Mechanics is often needed to describe the physical properties of atoms and molecules properly. What makes it even more complicated is that nanostructured materials can still be quite a challenge to describe Quantum Mechanically, since the equations are feindishly difficult - this is where supercomputers come in to make the job approachable. The scientific discipline seeking to increase our knowledge and understanding of nanoscale phenomena, i.e. science on the scale of 0.1 nm to 100 nm.

Nanosensor: A chemical or physical sensor constructed using nanoscale components, usually microscopic or submicroscopic in size.

Nanoshell: A nanoparticle composed of a metallic shell surrounding a semiconductor. When nanoshells reach a target cancer cell, they can be irradiated with near-infrared light or excited with a magnetic field, either of which will cause the nanoshell to become hot, killing the cancer cell.

Nanoshells: Nanoscale metal spheres, which can absorb or scatter light at virtually any wavelength. "The nanoshells act as an amazingly versatile optical component on the nanometer scale: they may provide a whole new approach to optical materials and components," Professor Naomi Halas. See Metal Nanoshells in Bioengineering and Nanoshells May Be Key To Next Wave Of Light-Based Technology and Physics of Nanoshells.

Nanosieving: In medical nanorobotics, a nanodevice that can sort molecules or other nanoscale objects by physical sieving.

Nanosieving: In medical nanorobotics, a nanodevice that can sort molecules or other nanoscale objects by physical sieving.

Nanosome: Nanodevices existing symbiotically inside biological cells, doing mechanosynthesis and disassembly for it and replicating with the cell. Similar to nanochondria. [AS January 1998]

Nanosome: Nanodevices existing symbiotically inside biological cells, doing mechanosynthesis and disassembly for it and replicating with the cell. Similar to nanochondria. [Anders Sandberg January 1998]

Nanosources: sources that emit light from nanometre-scale volumes.

Nanospheres: Self-assembling nanospheres that fit inside each other like Russian dolls are one form of a broad range of submicroscopic spheres ... The durable silica spheres, which range in size from 2 to 50 nanometers, form in a few seconds, are small enough to be introduced into the body, and have uniform pores that could enable controlled release of drugs. The spheres can absorb organic and inorganic substances including small particles of iron, which means they can be controlled by magnets and the contents released as needed. [Sandia National Labs news release "Self- assembled spheres may be helpful against disease or terrorism".

Nanosprings: A nanowire wrapped into a helix. Speculation is that they "may someday make highly sensitive magnetic field detectors, perhaps finding application in hard drive read heads. Alternatively, nanosprings could serve as positioners, or even as tiny conventional springs, for nanomachines of the future." See Spiraling in on Nanosprings and Nanosprings jump into place.

Nanostructures: Nanometer sized objects.

Nanostructures may be considered as small, familiar, or large, depending on the view point of the disciplines concerned. To chemists, nanostructures are molecular assemblies of atoms numbering from 10^3 to 10^9 and of molecular weights of 10^4 to 10^{10} Daltons. Thus, they are chemically large supramolecules. To molecular biologists, nanostructures have the size of familiar objects from proteins to viruses and cellular organelles. But to material scientists and electrical engineers, nanostructures are the current limit of microfabrication and thus are rather small. Nanostructures are complex systems which evidently lie at the interface between solid- state physics, supramolecular chemistry, and molecular biology.

Nanosurgery: A generic term including molecular repair and cell surgery.

Nanoswarm: UFog and Goo

Nanosystem: A eutactic set of nanoscale components working together to serve a set of purposes; complex nanosystems can be of macroscopic size.

Nanotech: Slang for nanotechnology.

Nanotechism: the religion of nanotech, as opposed to the science of nanotech

Nanotechnology Characterization Lab NCL Spotlight: Eliminating suffering and death from cancer requires an unprecedented collaborative effort that leverages resources from government, industry, and academia. Working in concert with the National Institute of

Standards and Technology (NIST) and the U.S. Food and Drug Administration (FDA), the National Cancer Institute (NCI) established the Nanotechnology Characterization Laboratory to perform preclinical efficacy and toxicity testing of nanoparticles.

The NCL will serve as a national resource and knowledge base for all cancer researchers to facilitate the regulatory review of nanotechnologies intended for cancer therapies and diagnostics. By providing the critical infrastructure and characterization services to nanomaterial providers, the NCL can accelerate the transition of basic nanoscale particles and devices into clinical applications, thereby reducing suffering and death from cancer.

As part of its assay cascade, the NCL will characterize nanoparticles' physical attributes, their *in vitro* biological properties, and their *in vivo* compatibility using animal models. The time required to characterize nanomaterials from receipt through the *in vivo* phase is anticipated to be one year.

Nanotechnology: A manufacturing technology able to inexpensively fabricate most structures consistent with natural law, and to do so with molecular precision. (2) In recent general usage, any technology related to features of nanometer scale: thin films, fine particles, chemical synthesis, advanced microlithography, and so forth. As introduced by the author, a technology based on the ability to build structures to complex, atomic specifications by means of mechanosynthesis; this can be termed molecular nanotechnology. (3) Nanotechnology is the knowledge of the processes needed to make useful things by controlling matter on the level of atoms and molecules. It is accepted now that the upper boundary of size in this field is 100 nm. Nanotechnology promises to create a new class of imaging agents that offer distinct advantages and can be used for the early detection and diagnosis of disease. Diagnostic molecules have the potential to act as biomarkers in drug development and diagnostics, and can be used in the imaging of cancer in living subjects. This meeting will address the challenges in implementing nanotechnology for drug delivery systems and imaging agents, and promote dialogue between diagnostic and therapeutic development. The production and application of structures, devices and systems by controlling shape and size at nanometre scale.

The development and use of techniques to study physical phenomena and construct structures in the nanoscale size range or smaller. The creation of functional materials, devices and systems through control of matter at the scale of 1 to 100 nanometers, and the exploitation of novel properties and phenomena at the same scale. Nanotechnology is emerging as a field critical for enabling essential breakthroughs that may have tremendous potential for affecting biomedicine. Moreover, nanotechnologies developed in the next several years may well form the foundation of significant commercial platforms. Emerging as a new field enabling the creation and application of materials, devices, and systems at atomic and molecular levels and the exploitation of novel properties that emerge at the nanometer scale. Many areas of biomedicine are expected to benefit from nanotechnology including **sensors** for use in the laboratory, the clinic, and within the human body; new formulations and routes for drug delivery;

and biocompatible, high- performance materials for use in implants. Examples of potential uses of nanotechnology in biomedicine include the early detection and treatment of disease and the development of "smart", rejection- resistant implants that will respond appropriately as the body's needs change. Although research in this field dates back to Richard P. Feynman's classic talk in 1959, the term nanotechnology was first coined by K. Eric Drexler in 1986 in the book *Engines of Creation*. In the popular press, the term nanotechnology is sometimes used to refer to any sub- micron process, including lithography. Because of this, many scientists are beginning to use the term *molecular nanotechnology* when talking about true nanotechnology at the molecular level.

Technology based on the manipulation of individual atoms and molecules to build structures to complex, atomic specifications. The application of nanoscience in order to control processes on the nanometer scale, i.e. between 0.1 nm and 100 nm. The interactions of cellular and molecular components and engineered materials—typically clusters of atoms, molecules, and molecular fragments—at the most elemental level of biology. Such nanoscale objects—typically, though not exclusively, with dimensions smaller than 100 nanometers—can be useful by themselves or as part of larger devices containing multiple nanoscale objects. *See* Molecular nanotechnology.

Nanoterrorism: using MNT derived nanites to do damage to people or places.

Nano-test-tubes: CNT's opened and filled with materials, and used to carry out chemical reactions. See The

Opening and Filling of Multi-Walled Carbon Nanotubes (MWTs) and The Opening and Filling of Single-Walled Carbon Nanotubes (SWTs). A High Resolution Transition Electron Micrograph (TEM) of Samarium Oxide Inside a Multi-Walled Carbon Nanotube.

A one dimensional fullerene (a convex cage of atoms with only hexagonal and/or pentagonal faces) with a cylindrical shape. Carbon nanotubes discovered in 1991 by Sumio Iijima resemble rolled up graphite, although they can not really be made that way. Depending on the direction that the tubes appear to have been rolled (quantified by the 'chiral vector'), they are known to act as conductors or semiconductors. Nanotubes are a proving to be useful as molecular components for nanotechnology.

Strictly speaking, any tube with nanoscale dimensions, but generally used to refer to carbon nanotubes (a commonly mentioned non-carbon variety is made of boron nitride), which are sheets of graphite rolled up to make a tube. The dimensions are variable (down to 0.4 nm in diameter) and you can also get nanotubes within nanotubes, leading to a distinction between multi-walled and single-walled nanotubes. Apart from remarkable tensile strength, nanotubes exhibit varying electrical properties (depending on the way the graphite structure spirals around the tube, and other factors), and can be insulating, semiconducting or conducting (metallic).

Nanotube: A one dimensional fullerene with a cylindrical shape.

Nanotubes: Nanometer-sized tubes composed of various substances including carbon (CARBON

NANOTUBES), boron nitride, or nickel vanadate. A one dimensional *fullerene* (a convex cage of atoms with only hexagonal and/ or pentagonal faces) with a cylindrical shape. Carbon nanotubes discovered in 1991 by Sumio Iijima resemble rolled up graphite, although they can not really be made that way. Depending on the direction that the tubes appear to have been rolled (quantified by the 'chiral vector'), they are known to act as conductors or semiconductors. Nanotubes are a proving to be useful as molecular components for nanotechnology. [about.com]

Nanowetting: how wetting behavior depends on nanoscale topography on a substrate.

Nanowire: A nanometer-scale wire made of metal atoms, silicon, or other materials that conduct electricity. Nanowires are built atom by atom on a solid surface, often as part of a microfluidic device. They can be coated with molecules such as antibodies that will bind to proteins and other substances of interest to researchers and clinicians. By the very nature of their nanoscale size, nanowires are incredibly sensitive to such binding events and respond by altering the electrical current flowing through them, and thus can form the basis of ultra sensitive molecular detectors.

Nanowires: "Semiconductor nanowires are one-dimensional structures, with unique electrical and optical properties, that are used as building blocks in nanoscale devices." See Nanowires within nanowires and Learning how to Fabricate Nanowire. "Striped or 'superlatticed' nanowires can function as transistors, LEDs (light-emitting diodes) and other optoelectronic devices, biochemical sensors, heat-pumping thermoelectric devices, or all of the above, along the same length of wire." See Nanowires Get Their Stripes. Molecular wires millions of times smaller in diameter than a human hair. Described in a paper appearing in the February 23, 2001 issue of the journal Science, these "nanowires," so called because they have dimensions on the order of a nanometer (a billionth of a meter), have high rates of electron transfer with very low resistance. "That means less impedance to the flow of current, with little or no loss of energy," says chemist John Smalley, the lead Brookhaven researcher on the study. [News release Brookhaven National Lab (US) "Scientists Investigate "Nanowires" With Very Low Resistance" Feb. 22, 2001]

Nastic structures: The Nastic Materials Program is exploring the development of a new class of active materials that can mimic the ability of plants to generate large strains while still performing a structural function. The impetus for this work is the desire to develop a highly controllable and reversible material system that can generate 10 Megapascals and 20 percent in blocked stress and free strain, respectively. The ultimate goal is the development of plant- inspired actuation systems that possess the power density of conventional hydraulic systems for application in military adaptive or morphing structures.

National Nanotechnology Initiative: US federal government agencies participating include the National Science Foundation, the Department of Defense, the National Institute of Health, NASA, and NIST. [National Nanotechnology Initiative website] http://www.nano.gov/

Naturophilia: An exclusive love of Nature, disdaining everything that is artificial or technological.

Navicyte: In medical nanorobotics, a mobile, mass-storage (nanorobotic) device, similar to a communicyte, that may be used to establish a navigational network inside the human body.

NBIC: **N**anotechnology, **B**iotechnology, **I**nformation Technology and **C**ognitive Science. See Converging Technologies for Improving Human Performance

NE^{3}LS: Nanotechnology's Ethical, Environmental, Economic, Legal, and Social Implications. From 'Mind the gap': science and ethics in nanotechnology. click here (requires free registration) [Anisa Mnyusiwalla, Abdallah S. Daar and Peter A. Singer 2003 Nanotechnology 14 R9-R13. 13 Feb 2003]

NEG: Someone who typically complains, moans, and whines, Someone practicing the opposite of dynamic optimism.

NEMS (NanoElectroMechanical Systems): Systems consisting of integrated electromechanical actuators of nanometer-scale dimensions. The time is ripe for a concerted exploration of nanoelectromechanical systems (NEMS) i.e. machines, sensors, computers and electronics that are on the nanoscale. Such efforts are under way in my group at Caltech, and in several others around the world. The potential payoffs are likely to be enormous and could benefit a diverse range of fields, from medicine and biotechnology to the foundations of quantum mechanics.

NEMs: NanoElectroMechanical systems. Devices based on the movement of nanometer-scale components.

Neologomania: The transhuman enthusiasm for neologisms, especially the creation of words with many prefixes. [Brian Manning Delany]

Neomorph: A transhuman or posthuman with a non-humanoid body [Greg Bear, *Eon*]

Neophile: One who welcomes the future and who enjoys change and evolution.

Neophilia: (From the Jargon File, with a slight hacker bias): The trait of being excited and pleased by novelty. Common among most hackers, SF fans, and members of several other connected leading-edge subcultures, including the pro-technology 'Whole Earth' wing of the ecology movement, space activists, many members of Mensa, and the Discordian/neo-pagan underground. All these groups overlap heavily and (where evidence is available) seem to share characteristic hacker tropisms for science fiction, music, and oriental food. The opposite tendency is 'neophobia'.

Neophobe: One who fears change and wants to abort technological and social transformation.

Nernst equation: An equation relating the potential of an electrochemical cell to the concentrations of the cell components: $E = Eo + RT/zF \ln C_1/C_2$ with z the charge exchanged at the electrode and C_1 and C_2 concentrations of two electro-active compounds.

Neroprosthesis: Implanted cybernetic brain augmentation.

Neural Simulation: Imitating the functions of a neural system—such as the brain—by simulating the function of each cell.

Neurocomputation: The study of how natural and artificial neural networks process information.

Neuron Star: A neutron star used as a basis for a densely packed mind (similar to a jupiter brain or a omegon) exploiting neutronium or quark matter for computation [Damien Broderick, 1996]

Neuron: A nerve cell, such as those found in the brain.

Neuronaut: A person who explores their own neural functioning and internal mentational processes by various means, including deep introspection and meditation, psychoactive drugs, mind machines, and neuroscientific understanding.

Neurosuspension: Cryonically suspending only the head or brain of a person.

NMOS: An acronym for n-*channel metal-oxide-semiconductor*, as in NMOS transistor and NMOS logic.

Nono: A prefix meaning ten to the minus ninth power, or one billionth.

Nootropic: A cognition-enhancing drug that has no significant side-effects.

NOR: NOT-OR — logic gate whose output is the negation of that of the OR gate.

NOT: Logic gate whose output is binary 1 when its input is 0, and whose output is a 0 when its input is a 1.

Now shock: Being shocked or confused by the rapid changes that has already taken place, a kind of future shock before the present. [Michael Rothschild 1995, "Cornucopia Or Black Hole?", *Upside*]

NRAM™ - Nanotube-based/Nonvolatile RAM, developed by Nantero, using proprietary concepts and methods derived from leading-edge research in nanotechnology.

n-type: Characterizes a semiconductor containing predominantly mobile electrons (also see"p-type").

Nuclear Magnetic Resonance (NMR): An analytical method that can detect subatomic and structural information of molecules by measuring the adsorption of radio-frequency electromagnetic radiation by nuclei under the influence of a magnetic field.

Nucleation: The initial stage in a phase transformation, evidenced by the formation of small particles (nuclei) of the new phase, which are capable of growing.

Nucleotide: A monomer of the nucleic acids composed of a five-carbon sugar, a nitrogen-containing base, and phosphoric acid. A small molecule composed of three parts: a nitrogen base (a purine or pyrimidine), a sugar (ribose or deoxyribose), and phosphate. Nucleotides serve as the building blocks of nucleic acids (DNA and RNA).

Nucleus: In biology, a structure in advanced cells that contains the chromosomes and apparatus to transcribe DNA into RNA. In physics, the small, dense core of an atom. The positively charged core of an atom, an object of ~0.00001 atomic diameters containing >99.9% of the atomic mass. Nuclear positions define atomic positions.

Nutraceutical: A foodstuff or parts of foodstuff that offers medical or health benefits related to the prevention and treatment of disease.

O

OEIS OptoElectronic Integrated Systems: research is currently focused on the development of methodologies for self-assembly of micron scale objects with the objective of fabricating optoelectronic integrated systems (OEIS) [RP25]. In an extension of our integrated microarray research, electronically addressable test chips, comprising n x n matrices of microelectrodes each bearing photolithographically defined oligonucleotides of programmed base sequence, have been developed as experimental platforms for the self- assembly and interconnection of oligonucleotide modified optoelectronic components, e.g., LEDs and photodetectors. In this process, DNA- modified optoelectronic components may be transported to the surface of a microelectrode using electrophoresis whereupon sequence specific oligonucleotide interactions direct component localisation and binding. [NMRC (Ireland) Nanotechnology Research, Scientific Report 1999] http://www.nmrc.ie/reports/1999/scientific/scinano.html

Ohm (Ω): Unit of resistance. One ohm is the electrical resistance between two points of a conductor when a constant potential difference of 1 volt, applied to these points, produces a current of 1 ampere in the conductor.

Ohmmeter: Tool for measuring electrical resistance.

Okazaki fragments: Short lengths of DNA produced on the lagging strand during DNA replication, discovered by R. Okazaki. They are rapidly joined by DNA ligase to form a continuous DNA strand.

OLED or Organic LED: OLED or Organic LED is not made of semiconductors. It's made from carbon-based molecules. That is the key science factor that leads to potentially eliminating LEDs' biggest drawback ñ size. The carbon-based molecules are much smaller. And according to a paper written by Dr. Uwe Hoffmann, Dr. Jutta Trube and Andreas Kl–ppel, entitled OLED - A bright new idea for flat panel displays "OLED is brighter, thinner, lighter, and faster than the normal liquid crystal (LCD) display in use today. They also need less power to run, offer higher contrast, look just as bright from all viewing angles and are - potentially - a lot cheaper to produce than LCD screens." See also LCD and LED. LCD, LED, and OLED definitions courtesy The San Francisco Consulting Group (SFCG)

OLED or Organic LED: Is not made of semiconductors. Itís made from carbon-based molecules. That is the key science factor that leads to potentially eliminating LEDsí biggest drawback ñ size. The carbon-based molecules are much smaller. And according to a paper written by Dr. Uwe Hoffmann, Dr. Jutta Trube and Andreas Kl–ppel, entitled OLED - A bright new idea for flat panel displays ìOLED is brighter, thinner, lighter, and faster than the normal liquid crystal (LCD) display in use today. They also need less power to run, offer higher contrast, look just as bright from all viewing angles and are - potentially - a lot cheaper to produce than LCD screens. LCD, LED, and OLED definitions courtesy The San Francisco Consulting Group (SFCG)

Olefin: An alkene.

Oligodendrocyte: Type of glial cell in the vertebrate central nervous system that forms a myelin sheath around axons.

Oligomer: Short polymer, usually consisting (in a cell) of amino acids (oligopeptides), sugars (oligosaccharides), or nucleotides (oligonucleotides). (From Greek *oligos*, few, little.)

Omega point: Also called the Quantum Omega Point Theory. A possible future state when intelligence controls the Universe totally, and the amount of information processed and stored goes asymptotically towards infinity. See Terminology From The Omega Point Theory List . [Origin: Teilhard de Chardin, The Phenomenon of Man. See also Barrow and Tipler, The Cosmological Anthropic Principle or Tipler's The Physics of Immortality for a more modern definition.]

Omitted reaction: A chemical reaction that fails by not occurring (see misreaction).

Oncogene: One of a large number of genes that can help make a cell cancerous. Typically, a mutant form of a normal gene (proto-oncogene) involved in the control of cell growth or division.

Oocyte: Developing egg; usually a large and immobile cell.

Oogenesis: Formation and maturation of oocytes in the ovary.

Op-amp: Operational Amplifier — semiconductor amplifier characterized by high gain and high internal resistance.

Operator: Short region of DNA in a bacterial chromosome that controls the transcription of an adjacent gene.

Operon: In a bacterial chromosome, a group of contiguous genes that are transcribed into a single mRNA molecule.

Optical Trapping: Optical trapping of small particles by forces exerted by laser radiation pressure has recently been introduced into the study of biological systems. The single- beam gradient trap (*optical tweezers*) employs a single strongly focused **laser** beam. In this case, particle size is much less than and the laser light exerts a force pulling the particle toward the high focus part of the beam. Such manipulation of micron- sized (particle size larger than) particles is also feasible. Manipulations of cells and intracellular organelles have extended to laser cutting (scissors) and to use of *two- photon* systems. A method for non- mechanical manipulation of small particles. Small transparent objects (such as glass-spheres or yeast cells) can be trapped using focused *laser* light. The method has many possible applications in various fields of physics and biology. The optical trap is based on the concept of *photon* momentum transfer. Photon momentum was proposed by Planck and used by Einstein to explain the photo-electric effect.

Optical Tunneling: A quantum mechanical phenomena resulting from photon delocalization that allows light to cross propagation barriers such as an interface.

Optical Tweezers: Developed at Bell Labs in the 1980s ... The method relies on the fact that light waves exert minute forces as well as transmitting energy. If an object is small enough - in this case about 5 micrometers across — it can be "trapped" or held still by a focused beam of laser light. When the laser's position changes, the trapped object moves too ... The new technique allows researchers to make measurements not possible with conventional methods and should make it easier to judge the effectiveness of inhibitors and other medicines at a variety of concentrations. The single- beam gradient trap, or the *optical tweezers*, is a unique tool for micro- manipulation, enabling its user to access tiny objects without any mechanical contact. It uses a focussed laser beam to trap and manipulate microscopic particles in the size range 100 nm - 100 mm. The trap consists of a single laser beam which attracts transparent dielectric particles towards its focal region. The laser beam is introduced into a conventional optical microscope, so that the same objective is used to view and trap particles immersed in a liquid.

OR: Logic gate whose output is a binary 1 if any of its inputs is a 1; zero otherwise.

Orbital Tower: Also known as a "space tether", "beanstalk" or "heavenly funicular". A cable in synchronous orbit, with one end anchored to the surface of the Earth, often with a small asteroid at the outer end to provide some extra tension and stability. Picture also a "space elevator". In theory, constructed of a diamondoid material, approximately 22,000 miles long, with one end in a stable orbit, and the other somewhere [probably] around the equator. Used frequently in science-fiction yarns, and may become a reality with the advent of mature MNT. Such an elevator would move freight and passengers into orbit at a cost per pound orders of magnitude less than current launches, with passenger safety comparable to train, plane, or subway trips. Becomes possible when we can mass-produce nanotubes, and make their length to fit.

Orbital: In the approximation that each electron in a molecule has a distinct, independent wave function, the spatial distribution of an electron wave function corresponds to a molecular orbital. These, in turn, can be approximated as sums of contributions from the orbitals characteristic of the isolated atoms. An electron added to a molecule—or, similarly, one excited to a higher-energy state within a molecule—would occupy a state with a different wave function from the rest; an unoccupied state of this kind corresponds to an unoccupied molecular orbital. Orbital-symmetry effects on reaction rates arise when a reaction requires overlap between two lobes of the orbitals on each of two reagents: if the algebraic signs of the wave functions in the facing lobes do not match, bond formation between those orbitals is prohibited.

Organic Moledule: A molecule containing carbon; the complex molecules in living systems are all organic molecules in this sense.

Organotypic: Despite their wide use, the physiological relevance of organotypic slices remains controversial. Such cultures are prepared at 5 days postnatal. Although some local circuitry remains intact, they develop subsequently in isolation from the animal and hence without plasticity due to experience. Development of synaptic connectivity and morphology might be expected to proceed differently under these conditions than in a behaving animal.

Oscillator: Circuit that produces an alternating voltage (current) when supplied by a steady (DC) energy source.

Osmosis: The flow of solvent into a solution through a semi-permeable membrane.

Osmotic pressure (ð): The pressure that must be applied to a solution to stop osmosis: ð = MRT.

Outmessaging: In medical nanorobotics, conveyance of information from a transmitter located inside the human body, especially from working nanodevices, to the patient or to a recipient external to the human body.

Overcoming Cytoxicity: Studies have shown that quantum dot toxicity is due to cadmium ions being released into cells. This is exacerbated by oxygen, UV exposure and the large exposed surfaces of the spherical quantum dots. To avoid these problems, surfaces are added to the quantum dots, traditionally consisting of Zinc-Sulfide compounds to create a more benign surface. Polymers are used as well, which have been shown to be stable, but the polymers themselves may kill cells.

Overlap repulsion: A repulsive force resulting from the nonbonding overlap of two atoms.

Overlap: Orbitals lack sharply defined surfaces, declining in amplitude exponentially in their surface regions. When two orbitals are brought together, regions of substantial amplitude overlap. The resulting system can be described as two new orbitals, one formed by joining the two original orbitals without introducing a node in the wave function, and the other formed with a node between them. The nodeless joining reduces the energy of the electrons relative to the separate orbitals, resulting in a bonding interaction; joining with a node raises the energy, producing an antibonding interaction. If both new orbitals are occupied, antibonding forces dominate, resulting in overlap repulsion. Molecular mechanics models give an approximate description of overlap (and other) forces for a certain range of atoms and geometries.

Oxidation: The removal of one or more electrons from an atom, ion or molecule.

P

Package: Protective enclosure for a chip or a sensor, typically made of plastic or ceramic.

Palindromic sequence: Nucleotide sequence that is identical to its complementary strand when each is read in the same chemical direction - for example, GATC.

Paradigm Shift: When one conceptual world-view is replaced by another, or, a change of patterns on a massive scale. When Copernicus showed how the Earth rotates around the Sun, and not vice versa, that created a paradigm shift [it forced a new way of thinking about our place in the Universe]. And when quantum physics and general relativity displaced Newtonian mechanics, that created another shift. Applied to an enabling technology such as molecular manufacturing, it suggests that there will be many shifts occurring, soon, and with wide-ranging and often disruptive consequences. For more detail, see Accelerating Intelligence: Where Will Technology Lead Us? [by Ray Kurzweil].

Paramagnetism: A relatively weak form of magnetism resulting from the independent alignment of atomic dipoles (magnetic) with an applied magnetic field.

Also a type of induced magnetism, associated with unpaired electrons, that causes a substance to be zapped into the inducing magnetic field.

Partition function: A function determined by the probability distribution (over phase space in the classical treatment; over quantum states in the quantum treatment) describing a thermally equilibrated system; many thermodynamic quantities can be expressed in terms of the partition function and its derivatives.

Patch-clamp recording: Electrophysiological technique in which a tiny electrode tip is sealed onto a patch of cell membrane, thereby making it possible to record the flow of current through individual ion channels in the patch.

Pathogen (adjective pathogenic): An organism or other agent that causes diseases.

PCR (polymerase chain reaction): Technique for amplifying specific regions of DNA by multiple cycles of DNA polymerization, each followed by a brief heat treatment to separate complementary strands.

PDF: See probability density function.

Peptide bond: Chemical bond between the carbonyl group of one amino acid and the amino group of a second amino acid - a special form of amide linkage.

Peptide map: Characteristic two-dimensional pattern (on paper or gel) formed by the separation of the mixture of peptides produced by the partial digestion of a protein.

Peptide nanotubes: NANOTUBES formed from cyclic peptides (PEPTIDES, CYCLIC). Alternating D and L linkages create planar rings that self assemble by stacking into nanotubes. They can form pores through CELL MEMBRANE causing damage. MeSH 2004

Peptide: A short chain of amino acids; see protein.

Permeability [μ]: From the relation between magnetic induction and magnetic field (B = μH); for free space, $\mu_0 = 1.26 \times 10^{-6}$ H/m.

Permitivity [å]: From the relation between polarization charge and electric field; for free space, $å_0 = 8.85 \times 10^{-12}$ F/m.

Peroxisome: Small membrane-bounded organelle that uses molecular oxygen to oxidize organic molecules. Contains some enzymes that produce and others that degrade hydrogen peroxide (H2O2)

Pervasive Computing: When computers (and sensors and actuators) become virtually invisible, and are used in almost every aspect of human commerce, interaction, and life. It will allow you full control over data and information, enabling you to send, receive, manage, and update your data from anywhere at any time. It will also allow you full control over your environment, in so far as you will be able to speak or gesture commands, effecting changes to things around you. Applications include: environmental monitoring - when you enter a room, they sense your presence and adjust temperature and humidity to your personal preferences; building security - to sense chemical weapons and perform face recognition; information transfer - allowing you to send and receive calls, data, and images from anywhere to anywhere, without the need of bulky equipment. Also called "Ubiquitous

Computing", "Intelligent Telesensing", "Proactive Computing", "Distributed Information Management Systems", "The Evernet", and "Calm Technology". "*...it will look like nothing to the naked eye. ...beneath the surface, tiny computing networks will be doing exactly what we want them to do - working behind the scenes to help us see clearer, travel safer, and place more knowledge, rather than frustration, into our heads.*" [Howard Lovy, editor at Small Times Media]

PES: *See* potential energy surface.

pH: Common measure of the acidity of a solution: "p" refers to power of 10, "H" to hydrogen. Defined as the negative logarithm of the hydrogen ion concentration in moles per liter (M). Thus pH 3 (10E-3 M H+) is acidic and pH 9 (10E-9 M H+) is alkaline.

Phage: Any virus that infects bacteria. They were the first organisms used for the study of molecular genetics and are now widely used as cloning vectors. (From Greek *phagein*, to eat.)

Phagocyte: General term for a professional phagocytic cell - that is, a cell such as a macrophage or neutrophil that is specialized to take up particles and microorganisms by phagocytosis.

Phagocytosis: Process by which particulate material is endocytosed ("eaten") by a cell. Prominent in carnivorous cells, such as *Amoeba proteus*, and in vertebrate macrophages and neutrophils. (From Greek *phagein*, to eat.)

Pharmacyte: In medical nanorobotics, a theorized (nanorobotic) device capable of delivering precise doses of biologically active chemicals to individually-addressed human body tissue cells (e.g. cell-by-cell drug delivery).

Phase shift: A time difference between the input and output signals.

Phase space: A classical system of N particles can be described by its $3N$ position and $3N$ momentum coordinates. The phase space associated with the system is the $6N$ dimensional space defined by these coordinates.

Phase transformation: A change in the number and/or character of the phases that make up the microstructure of an alloy.

Phonon: A particle of sound. The energy E of a phonon is given by the Einstein relation, $E = hf$. Here f is the frequency of the sound and h is Planck's constant. The momentum p of a photon is given by the de Broglie relation, $p = h/ë$. Here ë is the wavelength of the sound. A quantum of acoustic energy, analogous to the quantum of electromagnetic radiation, the photon. Thermal excitations in a crystal or in an elastic continuum can be described as a population of phonons (analogous to blackbody electromagnetic radiation). In highly inhomogeneous solids, a description in terms of phonons breaks down and localized vibrational modes become important. A single quantum of vibrational or elastic energy.

Phosphorescence: Luminescence that occurs at times greater than on the order of a second after an electron excitation event (see also"luminescence").

Photodiode: Semiconductor diode that produces voltage (current) in response to illumination (see also"phototransitor").

Photolithographic Mask: A template used in photolithography that allows selective exposure of a photosensitive surface.

Photolithography: Writing or creating patterns by means of light.

Photomicrograph: The picture made with a microscope.

Photon: A particle of light. The energy E of a photon is given by the Einstein relation, $E = hf$. Here f is the frequency of the light and h is Planck's constant. The momentum p of a photon is given by the de Broglie relation, $p = h/ë$. Here ë is the wavelength of the light.

Photonic Band Gap (PBG): An energy range (and corresponding wavelength range) for which a material neither absorbs light nor allows light propagation.

Phototransistor: Transistor that, when powered, produces amplified voltage (current) in response to illumination.

Pi bond: A covalent bond formed by overlap between two *p* orbitals on different atoms (see *sp*). Pi bonds are superimposed on sigma bonds, forming double or triple bonds.

Pico Technology: (trillionth of a meter) — The next step smaller, after Nano-technology. The art of manipulating materials on a quantum scale.

Piezoelectric material: A ferroelectric material in which an electrical potential difference is created due to mechanical deformation, or conversely, in which the application of a voltage causes dimensional changes in the material.

Piezoelectric: A piezoelectric material expands or contracts when a voltage is applied across it. Piezoelectric tubes are often used to move the tip in an AFM or STM. If a piezoelectric crystal is compressed, a voltage appears across it. This effect is sometimes used to measure displacements. A commonly used piezoelectric is PZT (Pb,Zr)TiO_3.

Pinhole: The term pinhole embraces a wide variety of oxide defects and is used in a broad sense today. Listed in this category are cracks caused by thermal contraction after oxidation or by handling, and regions of oxide with low dielectric strength caused by dust particles, inadequate masking, contamination, or poor resist adhesion.

Pink Goo: (humorous) Humans (in analogy with grey goo). "Pink Goo to refer to Old Testament apes who see their purpose as being fruitful and multiplying, filling up of the cosmos with lots more such apes, unmodified." [Eric Watt Forste August 1997]

Pin-out: Diagram showing for electronic components the relations between connecting pins and internal components.

Pitting: A form of very localized corrosion wherein small pits or holes form, usually in a vertical direction.

Pixel: Picture Element — smallest element of an image, such as a dot on a computer monitor screen.

pK value: A measure of the strength of an acid on a logarithmic scale. The pK value is given by log10 (1/Ka), where Ka is the acid dissociation constant pK values often are used to compare the strengths of different acids.

Plastic deformation: Permanent or nonrecoverable deformation, accompanied by permanent atomic displacement.

Plasticizer: A low molecular weight polymer additive that enhances flexibility and workability and reduces stiffness and brittleness.

Point defect: A crystalline defect associated with one or several atomic sites.

Poisson's ratio (í): For elastic deformation, the negative ratio of lateral and axial strains that result from an applied axial stress.

Poisson's ratio: A bar of an isotropic, elastic material ordinarily shrinks laterally when it is stretched longitudinally. The lateral contracting strain divided by the applied tensile strain is Poisson's ratio, which varies from material to material.

Polarization (P): The total electric dipole moment per unit volume of dielectric material.

Polyclonal antibodies: Antibodies produced by an animal's white blood cells (lymphocytes, specifically) in response to an antigen. This response occurs naturally or can purposely be created by injecting an animal, such as a rabbit or goat, with a specific antigen. More than one kind of anti-body is produced since more than one lymphocyte is producing antibodies. This is referred to as"polyclonal". The polyclonal antibodies are isolated from the animal and can be used for detection purposes. Because the antibodies are actually a mixture with different affinities (binding capability) for the antigen of interest, some variability in performance can occur from one test to another or one batch of antibodies to another.

Polycyclic: A cyclic structure contains rings of bonds; a structure having many such rings is termed polycyclic. In the polycyclic structures of interest in this volume, a large fraction of the atoms are members of multiple small rings, resulting in considerable rigidity.

Polymer: A molecule consisting of many smaller subunits covalently linked together.

Polymerase Chain Reaction (PCR): A chemical reaction that uses the polymerase enzyme to carry out *in vitro* replication of DNA.

Polymers - biomedical: More and more therapeutic problems are relevant to the use of polymer- based therapeutic aids for a limited period of time, namely the healing time related to the outstanding capacity of living systems to self-repair, e.g. bone fracture fixation with screws and plates, of wound closure by sutures and also of drug delivery from implants or similar systems based on polymeric matrices, or on aqueous dispersions or solutions of polymers. After healing the remaining prosthetic materials or devices become foreign residues or wastes that have to be eliminated from the body. Nowadays, biocompatible polymers that can degrade in the body are developed. The degradation and the elimination of degradation by-products depend on rather complex phenomena that are presently reflected inconsistently by terms issued from the tradition because each domain has developed its own terminology almost independently. This is a source of misunderstandings, confusions and misperceptions among scientists, surgeons, pharmacists, journalists and politicians, the situation being increased by the introduction of degradable polymers in plastic waste management and environmental protection.

Polysilicon: Polycrystalline silicon used as conductor in integrated circuits, and especially FETs. Short for Polycrystalline Silicon, used in the manufacture of computer chips.

Positional Assembly: Constructing materials an atom or molecule at a time

Positional Controlled Chemical Synthesis or Positional Synthesis: Control of chemical reactions by precisely positioning the reactive molecules, the basic principle of assemblers

Positional Devices: *See* "A New Family of Six Degree Of Freedom Positional Devices" by Ralph C. Merkle

Positional navigation: In medical nanorobotics, a form of nanorobotic navigation in which nanodevices know their exact location inside the human body to ~micron accuracy continuously at all times.

Positional synthesis: Control of chemical reactions by precisely positioning the reactive molecules; the basic principle of assemblers.

Positive Sum: A term used to describe a situation in which one or more entities can gain without other entities suffering an equal loss - for example, a growing eccnomy.

POSS Nanotechnology: short for Polyhedral Oligomeric Silsesquioxanes Nanotechnology. POSS nanomaterials are attractive for missile and satellite launch rocket applications because they offer effective protection from collisions with space debris and the extreme thermal environments of deep space and atmospheric re-entry. Another application of

POSS nanotechnology under development is a new high-temperature lubricant. This new nanolubricant is effective at temperatures up to 500ƒF, which is 100ƒF greater than conventional lubricants. From Technologies developed by the Propulsion Directorate's Polymer Working Group at Edwards AFB

Post Monetary Economy: After the advent of mature Nanotechnology, it is likely that our economic reality will change, possibly to the extent of eliminating currency as we know it today. See "Proposal for an Ideal Nano-Specie: Gold-Pressed Latinum" Robert A. Freitas Jr.

Posthuman: Persons of unprecedented physical, intellectual, and psychological capacity, self-programming, self-constituting, potentially immortal, unlimited individuals.

Potential energy surface: The potential energy of a ground-state molecular system containing N atoms is a function of its geometry, defined by $3N$ spatial coordinates (a configuration space). If the energy is imagined as corresponding to a height in a $3N + 1$ dimensional space, the resulting landscape of hills, hollows, and valleys is the potential energy surface.

Potential Energy: Energy stored in an object due to its position or internal structure. The energy associated with a configuration of particles, as distinct from their motions. In macroscopic experience, potential energy can be increased (for example) by stretching a spring or by lifting a mass against a gravitational force; in molecular systems, potential energy can be increased (for example) by stretching a bond or by separating molecules against a van der Waals attraction.

Potential well: In a potential energy surface, the region surrounding a local energy minimum. Typically taken to include at least those points in configuration space such that a path of steadily declining energy can be found that leads to the minimum in question, and such that no similar path can be found to any other minimum. If the PES were a landscape, this would be the region around the minimum that could be filled with water without any flowing down and away toward another minimum.

Potentiometric Device: Monitors the voltage between a sensing electrode and a reference electrode. A high input impedance voltmeter is used to minimize current flow. The voltage typically is proportional to the logarithm of the analyte concentration.

Power [W]: Product of voltage and current in a component; also, refers to the field of electric energy supply.

Precision: The degree of reproducibility among several independent measurements of the same true value under specified conditions.

Presentation semaphore: In medical nanorobotics, a mechanical device used to display specific antigens, chemical ligands, or other molecular objects to the external environment, with the purpose of selectively modifying the chemical or other surface characteristics of a nanorobot exterior.

Printed circuit board: PCB — selectively metallized insulating sheet for supporting and interconnecting circuit components.

Probability density function: Consider an uncertain physical property and a corresponding space describing the range of values that the property can have (e.g., the configuration of a thermally excited N particle system and the corresponding $3N$ dimensional configuration space). The probability density function associated with a property is defined over the corresponding space; its value at a particular point is the probability per unit volume that the property has a value in an infinitesimal region around that point.

Protein Design, Protein Engineering: The design and construction of new proteins; an enabling technology for nanotechnology.

Protein Folding: "The process by which proteins acquire their functional, preordained, three-dimensional structure after they emerge, as linear polymers of amino acids, from the ribosome." [The Scientist]

Protein: Living cells contain many molecules that consist of amino acid polymers folded to form more-or-less definite three-dimensional structures; these are termed proteins. Short polymers lacking definite three-dimensional structures are termed peptides. Many proteins incorporate structures other than amino acids, either as covalently attached side chains or as bound ligands. Molecular objects made of protein form much of the molecular machinery of living cells.

Proteomics: The study of the full expression of proteins in an organism The term proteome refers to all the proteins expressed by a genome, and thus proteomics involves the identification of proteins in the body and the determination of their role in physiological and pathophysiological functions. ... Ultimately it is believed that through proteomics new disease markers and drug targets can be identified that will help design products to prevent, diagnose and treat disease. [e-proteomics.net]

Proximal probes: A family of devices capable of fine positional control and sensing, including scanning tunneling and atomic force microscopes; an enabling technology for nanotechnology.

p-Type semiconductor: A semiconductor for which the predominant charge carriers responsible for electrical conduction are holes.

Pyroelectricity: The property of certain crystals, such as tourmaline, of acquiring opposite electrical charges on opposite faces when heated.

Q

Q factor: A rating, applied to coils, capacitors, and resonant circuits, equal to the reactance divided by the resistance. The ratio of energy stored to energy dissipated per cycle in an electrical or mechanical system.

Quantization: The concept that energy can occur only in discrete units called quanta.

Quantum Computer: A computer that exploits quantum mechanical phenomena such as superposition and entanglement.

Quantum Confined Atoms (QCA): atoms caged inside nanocrystals. May find uses in clear-glass sunglasses, bio-sensors, and optical computing.

Quantum corral: Structures where electrons are corralled (confined) by atoms placed on the surface of a metal. See Don Eiger's STM gallery for examples.

Quantum Cryptography: A system based on quantum-mechanical principles. Eavesdroppers alter the quantum state of the system and so are detected. Developed by Brassard and Bennett, only small laboratory demonstrations have been made.

Quantum dot (Qdots): Nanometer sized semiconductor

particles, made of cadmium selenide (CdSe), cadmium sulfide (CdS) or cadmium telluride (CdTe) with an inert polymer coating. The semiconductor material used for the core is chosen based upon the emission wavelength range being targeted: CdS for UV-blue, CdSe for the bulk of the visible spectrum, CdTe for the far red and near-infrared, with the particle's size determining the exact color of a given quantum dot. The polymer coating safeguards cells from cadmium toxicity but also affords the opportunity to attach any variety targeting molecules, including monoclonal antibodies directed to tumor-specific biomarkers. Because of their small size, quantum dots can function as cell- and even molecule-specific markers that will not interfere with the normal workings of a cell. In addition, the availability of quantum dots of different colors provides a powerful tool for following the actions of multiple cells and molecules simultaneously. In August 2004, researchers announced the successful preparation of water-soluble gold quantum dots that can also be constructed to emit light at a variety of wavelengths. These polymer-coated quantum dots may prove to be more suitable for use in human clinical applications.

Quantum Dot Nanocrystals (QDNs): Used to tag biological molecules, and "measuring between five and ten nanometres across, are made up of three components. Their cores contain paired clusters of atoms such as cadmium and selenium that combine to create a semiconductor. This releases light of a specific colour when stimulated by ultraviolet of a wide range of frequencies. These clusters are surrounded by a shell made of an inorganic substance, to protect them. The whole thing is then coated with an organic surface, to allow the attachment of proteins or DNA molecules. By varying the number of atoms in the core, QDNs can be made to emit light of different colours." [From The Economist print edition]

Quantum Dot: An object so small that adding or removing a single electron represents a significant change. Nanometer sized fragments (the dots) of semiconductor crystalline material which emits PHOTONS. The wavelength is based on the quantum confinement size of the dot. They are brighter and more persistent than organic chemical INDICATORS. They can be embedded in MICROBEADS for high throughput ANALYTICAL CHEMISTRY. Do not confuse with microscopic fluorescent bar codes which are micrometer sized.

An important strategy for nonisotopic labeling of *single molecules* is the use of highly luminescent semiconductor *nanocrystals*, or 'quantum dots,' that can be covalently linked to biological molecules. This class of detectors, which range in size from 1-5 nm, have been exploited for biological labeling by a number of laboratories, particularly those of Shimon Weiss, Paul Alivisatos and Shuming Nie (4, 5). Quantum dots offer several advantages over organic *dyes*, including increased brightness, stability against photobleaching, a broad continuous excitation spectrum, and a narrow, tunable, symmetric emission spectrum.

Because quantum dots are nontoxic and can be made to dissolve in water, efforts are underway to explore their use in labeling single molecules in living cells. Nanometer-sized semiconductor crystals, or electrostatically confined electrons. Something (usually a semiconductor island) capable of confining a single electron, or a few, and in which the electrons occupy discrete energy states just as they would in an atom (quantum dots have been called "artificial atoms").

Other terminology reflects the preoccupations of different branches of research: microelectronics folks may refer to a "single-electron transistor" or "controlled potential barrier," whereas quantum physicists may speak of a "Coulomb island" or "zero-dimensional gas" and chemists may speak of a "colloidal nanoparticle" or "semiconductor nanocrystal."

All of these terms are, at various times, used interchangeably with "quantum dot," and they refer more or less to the same thing: a trap that confines electrons in all three dimensions. [from Hacking Matter: Levitating Chairs, Quantum Mirages, and the Infinite Weirdness of Programmable Atoms. Wil McCarthy. February 2003] See also Quantum Dot and Artificial material structure for an explanation. Materials Today June 2002. (PDF)

Quantum dots: Quantum dots are regions of semiconductors that can be occupied by a few electrons. These dots have many properties that are similar to atoms. More about quantum dots.

Quantum Interferometric Lithography: A method of lithography that exploits quantum entanglement in order to focus a lithographic beam beyond half of its wavelength.

Quantum Mechanics: A largely computational physical theory explaining the behavior of quantum phenomena, which incorporates the theory of special relativity. Despite dilignet attempts, general relativity has not been sucessfully incorporated into quantum mechanics.

A physical model of chemical and optical phenomena, as well as the behaviour of matter in general on a small scale. Quantum mechanics describes a system of particles in terms of a wave function defined over the configuration space of the system. Although the concept of particles having distinct locations is implicit in the potential energy function that determines the wave function (e.g., of a ground-state system), the observable dynamics of the system cannot be described in terms of the motion of such particles from point to point.

In describing the energies, distributions, and behaviors of electrons in nanometer-scale structures, quantum mechanical methods are necessary. Electron wave functions help determine the potential energy surface of a molecular system, which in turn is the basis for classical descriptions of molecular motion. Nanomechanical systems can almost always be described in terms of classical mechanics, with occasional quantum mechanical corrections applied within the framework of a classical model.

Quantum Mirage: A nanoscale property that may allow information to be transfered through use of the wave property of electrons. Thus, quantum computers might not require wires as we know them. A nanoscale property that may allow information to be transfered through use of the wave property of electrons.

Quantum physics: Describes fundamental electronic and optical properties of matter at microscopic level and wave and interference phenomena in particular.

Quantum electronics, quantum optics and optoelectronics are important areas of application - atomic clocks, lasers, light emitting diodes, optical fibers, tunnel diodes and superconducting systems are important and well-known examples. The trend towards faster and more complicated microprocessors and microelectronics has resulted in electronic components now approaching the domains of quantum physics at research level. In about 20 years time, miniaturization will also be halted in commercial applications. It will then probably utilize electronic wave phenomena and single- electron effects in semiconductor components and systems and also to create more complicated transistor components connected in more complicated ways where quantum phenomena, cooperative phenomena and even superconductivity may be important for function. "QNANO" - *quantum nanophysics* - will probably provide a target for research and development in semiconductor physics, molecular electronics and bioelectronics in the foreseeable future.

Quantum Tunneling: When electrons pass through a barrier, without overcoming it or breaking it down. See Quantum tunneling and this illustration

Quantum Well: A P-N-P junction in which the "N" layer is ~10 nm (where traditional physics leaves off and quantum effects take over) and an "electron trap" is created. "If one makes a heterostructure with sufficiently thin layers, quantum interference effects begin to appear prominently in the motion of the electrons. The simplest structure in which these may be observed is a quantum well, which simply consists of a thin layer of a narrower-gap semiconductor between thicker layers of a wider-gap material." See Center for Quantum Electronics U of Dallas

Quantum Wire: Another form of quantum dot, but unlike the single-dimension "dot," a quantum wire is confined only in two dimensions - that is it has "length," and allows the electrons to propagate in a "particle-like" fashion. Constructed typically on a semiconductor base, and (among other things) used to produce very intense laser beams, switchable up to multi-gigahertz per second.

Quantum: Describes a system of particles in terms of a wave function defined over the configuration of particles having distinct locations is implicit in the potential energy function that determines the wave function, the observable dynamics of the motion of such particles from point to point. In describing the energies, distributions and behaviours of electrons in nanometer-scale structures, quantum mechanical methods are necessary. Electron wave functions help determine the potential energy surface of a molecular system, which in turn is the basis for classical descriptions of molecular motion. Nanomechanical systems can almost always be described in terms of classical mechanics, with occasional quantum mechanical corrections applied within the framework of a classical model. [NTN]

QuantumBrain: [theoretical] Think of your brain. Now, think of your brain performing at vastly superior levels. Nanobots will become an as-needed addition to your existing neurons, extending your mental capacities further then you can probably now imagine.

Quaternary Structure: Three-dimensional relationship of the different polypeptide chains in a protein complex.

Qubit: The quantum computing analog to a bit. Qubits exhibit superposition. Thus, unlike normal bits, qubits can be both 1 and 0 at the same time.

R

Radiation damage: Chemical changes (bond breakage, ionization) caused by high-energy radiation (e.g., x-rays, gamma rays, high-speed electrons, protons, etc.).

Radical: A structure with an unpaired electron (but excluding certain metal ions). In organic molecules, a radical is often associated with a highly reactive site of reduced valence (see doublet). The term radical is sometimes used to describe a substructure within a molecule; the term *free radical* then describes a radical in this sense, viewed as the result of cleaving the bond linking the substructure to the rest of the molecule.

Radioactive isotope: Form of an atom with an unstable nucleus that emits radiation as it decays.

Radioautography: *See* autoradiography

RAM: Random Access Memory — read-write memory with elements accessible in any order.

Range: The difference between the minimum and maximum values of sensor output in the intended operating range. Defines the overall operating limits of a sensor.

Rapid prototyper: a machine that can manufacture objects directly (usually, though not necessarily, in plastic) under the control of a computer.

Ras protein: One of a large family of GTP-binding proteins that help relay signals from cell-surface receptors to the nucleus. Named for the *ras* gene, first identified in viruses that cause rat sarcomas.

Reactance: Portion of impedance that characterizes non-dissipative, energy storage effects (also see"impedance").

Reactant: A starting substance in a chemical reaction. It appears to the left of the arrow in a chemical reaction.

Reaction: A process that transforms one or more chemical species into others. Typical reactions make or break bonds; others change the state of ionization or other properties taken to distinguish chemical species. In chemistry, any process in which the arrangement of atoms into molecules is changed.

Reading frame: The phase in which nucleotides are read in sets of three to encode a protein; an mRNA molecule can be read in any one of three reading frames.

Reagent device: A large reagent structure (or a large structure that binds a smaller reagent) serving as a component of a mechanochemical system. A reagent device exists chiefly to hold, position, and manipulate the environment of a reagent moiety.

Reagent moiety: The portion of a reagent device that is intimately involved in a chemical reaction.

Reagent: A chemical species that undergoes change as a result of a chemical reaction.

Receptor: A structure that can capture a molecule (often of a specific type in a specific orientation) owing to complementary surface shapes, charge distributions, and so forth, without forming a covalent bond. See dissociation constant.

Protein that binds a specific extracellular signaling molecule (ligand) and initiates a response in the cell. Cell-surface receptors, such as the acetylcholine receptor and the insulin receptor, are located in the plasma membrane, with their ligand-binding site exposed to the external medium. Intracellular receptors, such as steroid hormone receptors, bind ligands that diffuse into the cell across the plasma membrane.

Recessive: Refers to the member of a pair of alleles that fails to be expressed in the phenotype of the organism when the dominant member is present. Also refers to the phenotype of an individual that has only the recessive allele.

Recombinant DNA: Any DNA molecule formed by joining DNA segments from different sources. Recombinant DNAs are widely used in the cloning of genes, in the genetic modification of organisms, and in molecular biology generally.

Recombination: Process in which chromosomes or DNA molecules are broken and the fragments are rejoined in new combinations. Can occur in the living cell - for example, through crossing-over during meiosis - or in the test tube using purified DNA and enzymes that break and ligate DNA strands.

Reconstruction: A crystal consists of a regular array of atoms, and the simplest model of a crystal surface would be generated by simply discarding all atoms to one side of a surface without changing the positions of the rest. In reality, however, the positions of the remaining atoms do change. A pattern of displacements that lowers the symmetry of the surface (relative to the ideally terminated crystal) is termed a surface reconstruction; some reconstructions alter the pattern of bonds.

Recrystallization: The formation of a new set of strain-free grains within a previously cold-worked material due to an annealing heat treatment.

Rectifier: Device that converts bi-directional to one-way current flow.

Red blood cell: *See* erythrocyte

Red Goo: Deliberately designed and released destructive nanotechnology, as opposed to accidentally created grey goo.

Redox Group: A part of a molecule that undergoes changes in its electronic properties.

Reduced mass: Many dynamical properties of a system consisting of two interacting masses, m_1 and m_2, are equivalent to those of a system in which one mass is fixed in space and the other has a mass (the reduced mass) with the value $m_1m_2/(m_1 + m_2)$. The reduced mass description has fewer dynamical variables.

Reduction (verb reduce): Addition of electron density to an atom, as occurs during the addition of hydrogen to a molecule or the removal of oxygen from it. Opposite of oxidation.

Reduction: The addition of one or more electrons to an atom, ion, or molecule.

Redunancy: The use of more components than are needed to perform a function; this can enable a system to operate properly despite failed components.

Reflection: Deflection of a light beam at the interface between 2 media.

Refraction: Bending of a light beam when passing from one medium to another, at different velocities of light.

Refractory: A metal or ceramic that does not deteriorate rapidly or does not melt when exposed to extremely high temperatures.

Regenerative medicine: A field of medicine concerned with developing and using strategies aimed at repair or replacement of damaged, diseased, or metabolically deficient organs, tissues, and cells via TISSUE ENGINEERING; CELL TRANSPLANTATION; and ARTIFICIAL ORGANS and BIOARTIFICIAL ORGANS and tissues. MeSH 2004

Register: A temporary storage location for an array of bits within a digital logic system.

Relative magnetic permeability (μ_r): The ratio of the magnetic permeability of some medium to that of a vacuum (unitless), or: $\mu_r = \mu/\mu_0$, where μ_0 is the permeability of vacuum, a universal constant, which has a value of 1.257×10^{-6} H/m.

Relaxation time: A measure of the rate at which a disequilibrium distribution decays toward an equilibrium distribution. The electron relaxation time in a metal, for example, describes the time required for a disequilibrium distribution of electron momenta (e.g., in a flowing current) to decay toward equilibrium in the absence of an ongoing driving force and can be interpreted as the mean time between scattering events for a given electron.

Reliability (life, multi-use vs. single, calibration vs. accuracy drift): How well a sensor maintains both precision and accuracy over its expected lifetime. Also includes the robustness of the sensor.

Repeatability: The exactness with which a measuring instrument repeats indications when it measures the same property under the same conditions.

Replication: The mechanism used in living systems to make copies of genetic information.

Replicator: A system able to build copies of itself when provided with raw materials and energy. Any system that can build copies of itself when provided with the appropriate raw materials and energy.

In discussions of evolution, a replicator is an entity (such as a gene, a meme, or the contents of a computer memory disk) which can get itself copied, including any changes it may have undergone. In a broader sense, a replicator is a system which can make a copy of itself, not necessarily copying any changes it may have undergone. A rabbit's genes are replicators in the first sense (a change in a gene can be inherited); the rabbit itself is a replicator only in the second sense (a notch made in its ear can't be inherited).

Representative point: The point in a configuration space that represents the geometry of a system.

Reproducibility: A property of an experiment or process where one tends to receive consistent results from following a specific procedure.

Residual stress: A persisting stress in a material free of external forces or temperature gradients.

Resist: A material or coating that can protect a surface from chemical reactions.

Resistance [Ω-ohm]: Characteristic of a resistor: in a 1-ohm resistance a current of 1 A produces a voltage drop of 1 V.

Resistivity (ñ): The reciprocal of electrical conductivity, and a measure of a material's resistance to passing electric current.

Resistor: Energy dissipative element consisting of a poor conductor in series with connecting wires.

Resolution: The minimum distance between two objects that can be distinguished in microscopy or the minimum spacing between two features that can be fabricated with lithography. The smallest measurable change in input that will produce a small but noticeable change in the output. In the context of chemical separations, defines the completeness of separation.

Resonant frequency: The frequency at which a moving member or a circuit has a maximum output for a given input.

Respirocrit: In medical nanorobotics, the volume-fraction or bloodstream concentration of respirocyte nanorobots, expressed as a percentage.

Respirocyte: In medical nanorobotics, a theorized bloodborne spherical 1-micron (nanorobotic) device having a 1000-atm pressure vessel with active pumping powered by endogenous serum glucose, that serves as a mechanical artificial red blood cell.

Response time: The time it takes for the sensor's output to reach its final value. A measure of how quickly the sensor will respond to changes in the environment. In general, this parameter is a measure of the speed of the sensor and must be compared with the speed of the process.

Restriction Enzyme: An enzyme that cuts DNA at a specific site, allowing biologists to insert or delete genetic material.

Reverse bias: The insulating bias for a p-n junction rectifier; electrons flow into the p side of the junction.

RF: Radio Frequency — refers to alternating voltages and currents having frequencies between 9 kHz and 3 MHz.

Ribonuclease: An enzyme that cuts RNA molecules into smaller pieces.

Ribosome: A molecular machine, found in all cells, which builds protein molecules according to instructions read from RNA molecules. Ribosomes are complex structures built of protein and RNA molecules. A naturally occurring molecular machine that manufactures proteins according to instructions derived from the cell's genes.

Rigid structure: As used in this volume, a covalent structure that is reasonably stiff. In a typical rigid structure, all modes of deformation encounter first-order restoring forces resulting from some combination of bond stretching and angle bending; such a structure cannot undergo deformation by bond torsion alone. Meeting this condition usually requires a polycyclic diamondoid structure.

rms: Root mean square.

RNA: Ribonucleic acid; a molecule similar to DNA. In cells, the information in DNA is transcribed to RNA, which in turn is "read" to direct protein construction. Some viruses use RNA as their genetic material.

Roadmap: The semiconductor industry publishes a estimate of when new technologies will become available for use by the microelectronics industry.

ROM: Read only memory; memory used for permanent, storage of unalterable data; nonvolatile memory.

RNA: Ribonucleic acid; a molecule similar to DNA. In volve, the information in DNA is transcribed to RNA, which in turn is "read" to direct protein construction. Some viruses use RNA as their genetic material.

Roadmap: The semiconductor industry publishes estimate of when new technologies will become available for use by the microelectronics industry.

ROM: Read only memory; memory used for permanent storage of unalterable data; nonvolatile memory.

S

S phase: Period of a eucaryotic cell cycle in which DNA is synthesized.

***Saccharomyces*:** Genus of yeasts that reproduce asexually by budding or sexually by conjugation. Economically important in brewing and baking, they are also widely used in genetic engineering and as simple model organisms in the study of eucaryotic cell biology.

Sacrificial anode: An active metal or alloy that corrodes and protects another metal or alloy to which it is electrically coupled.

Sacrificial layer: A thin film that is later removed to release a microstructure from its substrate.

Salmonella: Rod-shaped, motile, aerobic genus of bacteria. Includes species that cause food poisoning.

Salt bridge: An ionic bond between charged groups that are part of larger covalent structures; salt bridges occur in many proteins.

SAMFET: (self assembled monolayer field effect transistor). Where a few molecules act as FETs, exhibiting both very strong gain, and extraordinarily rapid response. [Mark Ratner & MT 5(2) p. 20]

Sanguinatation: In medical nanorobotics, locomotion (especially swimming by a nanorobot) through the bloodstream.

Sapphirophagy: Eating sapphire (corundum).

Sarcoma: Cancer of connective tissue.

Sarcomere: Repeating unit of a myofibril in a muscle cell, composed of an array of overlapping thick (myosin) and thin (actin) filaments between two adjacent Z discs.

Sarcoplasmic reticulum: Network of internal membranes in the cytoplasm of a muscle cell that contains high concentrations of sequestered Ca2+ : that is released into the cytosol during muscle excitation.

Satellite DNA: Regions of highly repetitive DNA from a eucaryotic chromosome, usually identifiable by its unusual nucleotide composition. Satellite DNA is not transcribed and has no known function.

Saturated molecule: Molecule containing carbon-carbon bonds that has only single covalent bonds.

Saturated: An organic molecule is described as saturated if it is a closed shell species lacking double or triple bonds; forming a new bond to a saturated molecule requires the cleavage of an existing bond.

Scanning Capacitance Microscopy (SCM): A variation of scanning force microscopy that maps the localized capacitance of a surface. A method for mapping the local capacitance of a surface. [NTN]

Scanning electron microscope (SEM): A microscope producing an image by using reflected electron beams that scan the surface of a specimen.

Scanning Electron Microscopy (SEM): A method of producing images of a surface by scanning an electron beam over the sample and measuring the electronic interactions with the interface. *See* Virtual Scanning Electron Microscopy [Fl St U]

Scanning Force Microscope (SFM): An instrument able to image surfaces to molecular accuracy by mechanically probing their surface contours. A kind of proximal probe. .... A device in which the deflection of a sharp stylus mounted on a soft spring is monitored as the stylus is moved across a surface. If the deflection is kept constant by moving the surface up and down by measured increments, the result (under favorable conditions) is an atomic-resolution topographic map of the surface. Also termed an atomic force microscope.

Scanning Near Field Optical Microscopy: A method for observing local optical properties of a surface that can be smaller than the wavelength of the light used. A type of scanning probe microscopy that maps the near-field optical properties of an interface.

Scanning Probe Microscopy (SPM): A method for imaging nanoscale features of surfaces by scanning a sensor (probe) over a surface. Near-field effects such as tunneling, van der Waals forces, local fields and more are serially detected at localized points on the surface and used to create an SPM image.

Scanning Thermal Microscopy (SThM): A type of scanning probe microscopy that maps the local temperature and thermal conductivity of an interface. A method for observing local temperatures and temperature gradients on a surface. [NTN]

Scanning Tunneling Microscope (STM): An instrument able to image conducting surfaces to atomic accuracy; has been used to pin: molecules to a surface. A device in which a sharp conductive tip is moved across a conductive surface close enough to permit a substantial tunneling current (typically a nanometer or less). In a common mode of operation, the voltage is kept constant and the current is monitored and kept constant by controlling the height of the tip above the surface; the result, under favorable conditions, is an atomic-resolution map of the surface reflecting a combination of topography and electronic properties. The STM has been used to manipulate atoms and molecules on surfaces. A type of scanning probe microscopy that maps the local electrical properties of an interface.

Schwann cell: Glial cell responsible for forming myelin sheaths in the peripheral nervous system.

Science Court: A name (originally applied by the media) for a government-conducted fact forum.

Science: The process of developing a systematized knowledge of the world through the variation and testing of hypotheses. The pursuit of knowledge and understanding, from the Latin term *scientia*, which means 'knowledge'.

Sealed assembler lab: A general-purpose assembler system in a container permitting only energy and information to be exchanged with the environment.

Sealed Assembler Laboratory: A work space, containing assemblers, encapsulated in a way that allows information to flow in and out but does not allow the escape of assemblers or their products.

Second messenger: Small molecule that is formed in or released into the cytosol in response to an extracellular signal and helps to relay the signal to the interior of the cell. Examples include cAMP, IP3 and Ca2+.

Secondary structure: Regular local folding pattern of a polymeric molecule; in proteins, a helices and b-pleated sheets.

Secretory vesicle: Membrane-bounded organelle in which molecules destined for secretion are stored prior to release. Sometimes called secretory granule because darkly staining contents make the organelle visible as a small solid object.

Selectivity: The ability of a sensor to measure only one metric or, in the case of a chemical sensor, to measure only a single chemical species.

Self-Assembled Monolayer (SAM): A two-dimensional film, one molecule thick, covalently assembled at an interface.

Self-Assembler: A specific type of assembler that makes use of self-assembly such that the assembly process would theoretically not require external energy or information input.

Self-assembling biomolecular materials: Examples of self- assembly include protein folding, the formation of liposomes, and the alignment of liquid crystals. While this type of equilibrium self- assembly is the central focus of this report, it is important to emphasize that much biological assembly is also driven by energy sources such as adenosine triphosphate (ATP), which power biomotors.

Self-assembling peptides: Using standard peptide chemistry, we are able to manufacture small oligopeptide fragments which self- assemble into nano- fibers on a scale identical to *in vivo* extracellular matrix and impossible to synthetically produce by other manufacturing techniques. This explains why most of the alternatives to our PuraMatrix are harvested from animal and human cadavers (bovine collagen, cartilage extract, intestinal submucosa, mouse tumor extract, bovine gelatin, cadaver tissue, etc.) with high processing costs, high quality variability, and disease transmission risks. Synthetic alternatives fail to match the extremely fine fibers, high water content, and neutral charge that is so important to cells growing, dividing, expressing proteins, and remodeling tissue both in and outside of the body.

Self-Assembly: A method of integration in which the components spontaneously assemble, typically by bouncing around in a solution or gas phase until a stable structure of minimum energy is reached. A process in which supramolecular hierarchical organization is established without external intervention.... The approaches used can be expected to fall into two general categories. The first involves directly mimicking biological systems or processes to produce materials with enhanced properties. An example of this approach is the use of molecular genetic techniques to produce polymers with unprecedentedly uniform molecular length. The second category involves studying how nature accomplishes a task or creates a structure with unusual properties, and then applying similar techniques in a completely different context or using completely different materials. In chemical solutions, self-assembly (also called Brownian assembly) results from the random motion of molecules and the affinity of their binding sites for one another. Also refers to the joining of complementary surfaces in nanomolecular interaction. [ZY] See MITRE Nanosystems Research Task: Self-Assembly of Nanosystems and Microsystems

Self-repair: Indicating ability to heal itself without outside intervention.

Self-replication: More accurately labeled "exponential replication," self-replication refers to the process of growth or replication involving doubling within a given period. [ZY] Examp[le: create one assembler. Program it to create another, and program that one likewise, etcetera, until you have a speficied amount [which is the important part — how to make them STOP]. Self replication is an effective route to truly low cost manufacturing. Our intuitions about self replicating systems, learned from the biological systems that surround us, are likely to seriously mislead us about the properties and characteristics of artificial self replicating systems designed for manufacturing purposes. Artificial systems able to make a wide range of non- biological products (like diamond) under programmatic control are likely to be more brittle and less adaptable in their response to changes in their environment than biological systems. At the same time, they should be simpler and easier to design. The complexity of such systems need not be excessive by present engineering standards. The act of objects making copies of themselves - similar to biological reproduction except that self-replicating objects should make exact copies of themselves.

Semiconductor: A pure semiconductor is a poor electrical conductor but when certain impurities are added, the conductivity can increase by orders of magnitude. Common semiconductors are silicon, germanium, and gallium-arsenide. Computer chips are usually made from thin wafers cut from large single crystals of silicon. The silicon is a badly conducting matrix in which better conducting regions are defined by adding impurity atoms (often called dopants). The conductivity of a semiconductor can be described by the formula ó = *ne*ì, where ó is the conductivity measured in units of Ohm^{-1} m^{-1}, n is the density of free charge carriers measured in m^{-3}, e is the electron charge -1.6022×10^{-19} C, and ì is the mobility of the semiconductor measured in m^{2} V^{-1} s^{-1}. In a doped semiconductor, the density of free charge carriers is typically equal to the density of dopant atoms that have been added. Thus, the conductivity depends on the number of dopant atoms. Each dopant atom contributes one charge carrier. Silicon is the second most common element in the earth's crust (the first most common is oxygen). Silicon is commonly found in sand and stone in the form of silicon-dioxide. A substance or object with conductive properties between those of a conductor and a insulator. Material whose conductivity, due to charges of both signs, is normally in the range between that of metals and insulators and in which the electric charge carrier density can be changed by external means. [IUPAC Compendium] Nonmetallic material, such as silicon, whose electrical conductivity is moderate and alterable by doping.

Sensitivity: The amount of change in a sensor's output in response to a change at a sensor's input over the sensor's entire range. Provides an indication of a sensor's ability to detect changes. For some sensors, the sensitivity is defined as the input parameter change required to produce a standardized output change.

Sentience Quotient: In the article "Xenopsychology" by Robert Freitas in Analog of April 1984 there is an interesting index called "Sentience quotient". It is based on: The sentience of an intelligence is roughly directly related to the amount of data it can process per unit time and inversely to the overall mass needed to do that processing. This would be something like baud/kilograms. And since that would rapidly turn into a real big number, base 10 logs are used. The "least sentient" would be one bit over the lifetime of the universe massing the entire known universe, or about -70. The "most sentient" is claimed to be +50. Homo sapiens are around +13, a Cray I is +9, a venus flytrap is a peak of +1 with plants generally -2. [AS]

SER: *See* smooth endoplasmic reticulum

Serine/threonine kinase: Protein kinase that phosphorylates serines or threonines on its target protein.

Sertoli cell: Supporting cell of the mammalian testis that surrounds and nourishes developing sperm cells.

Sex chromosome: Chromosome that may be present or absent, or present in a variable number of copies, according to the sex of the individual; in mammals, the X and Y chromosomes.

Shape Memory Alloys: (SMA's) are a unique class of alloys which are able to "remember" their shape and are able to return to that shape even after being bent. The ability is known as the shape memory effect. ... This property has lead to many uses of SMA from orthodontics and coffee makers to methods of controlling aircraft and protecting buildings from earthquake damage. ... The first SMA to be discovered and the most commonly used is called Nitinol. [Texas A&M SMART lab] See also Introduction to Shape Memory and Superelasticity and Shape Memory Alloy Database

Shear modulus: Shear stress divided by shear strain; has the units of force per unit area.

Shear: A shear deformation is one that displaces successive layers of a material transversely with respect to one another, like a crooked stack of cards. Shear is a dimensionless quantity measured by the ratio of the transverse displacement to the thickness over which it occurs.

SI units: International System of units based on the metric system and units derived from the metric system.

Sigma bond: A covalent bond in which overlap between two atomic orbitals (e.g., of *sp, sp2,* or *sp3* hybridization) produces a single bonding orbital in which the distribution of shared electrons has a roughly cylindrical symmetry about the axis linking the two atoms; see pi bond, single bond, double bond, triple bond. By themselves, sigma bonds present little barrier to rotation of one substructure with respect to another, although steric effects and cyclic structures may hinder or block rotation.

Signal-to-noise-ratio: The ratio of the output signal with an input signal to the output signal with no input signal.

Single bond: A sigma bond having no associated pi bonds.

Single Cell Repair Unit: A cell repair unit using cilia for propulsion and equipped with a nanocomputer having 10 megabytes of fast RAM and 1 gigabyte of slower-access memory. The unit is extending 1000 individually-controlled molecular manipulators.

Single crystal: A crystalline solid for which the periodic and repeated atomic pattern extends throughout its entirety without interruption.

Single Electron Devices SED: Nanoscale devices that control the movement of individual electrons, may one day make it possible for integrated circuits to have as many as 10 billion electronic devices in a square centimeter, a density 1000 times greater than that believed feasible for conventional integrated circuits. In development since the mid-1980s, these devices consist of two electrodes (typically 30 nm wide) separated by a 1 nm- deep insulating layer through which single electrons can tunnel. These devices have many potential applications, from building more sensitive measurement devices to understanding fundamental problems in physics. In the last several years, researchers have built two- junction devices that share a middle electrode. These devices are called "*single- electron transistors*," because, like conventional transistors, their current can be controlled by modifying the surface charge on the middle electrode, making it an ideal element for an integrated circuit. A circuit made of single- electron devices, however, would have to be operated at a temperature of 4 K or below to reduce thermal effects which disturb the movements of single electrons in the solid.

Single Electron Transfer: Exchange of one electron at a time between two electrodes.

Single Foglet: "Imagine a microscopic robot. It has a body about the size of a human cell and 12 arms sticking out in all directions. A bucketfull of such robots might form a 'robot crystal' by linking their arms up into a lattice structure. Now take a room, with people, furniture, and other objects in it — it's still mostly empty air. Fill the air completely full of robots. The robots are called Foglets and the substance they form is Utility Fog, which may have many useful medical applications. And when a number of utility foglets hold hands with their neighbors, they form a reconfigurable array of 'smart matter.'" Copyright Dr. J. Storrs Hall Research Fellow of the Institute for Molecular Manufacturing. "Here's a short list of the powers you'd have or appear to have if embedded in fog: Creation—causing objects to appear and disappear on command. Levitation—causing objects to hover and fly around. Manipulation—causing forces (squeezing, hitting, pulling) on objects (real ones) at a distance. Teleportation—nearly any combination of telepresence and virtual reality between fog-filled locations." [Dr. J. Storrs Hall]

Singlet: An electronic state of a molecule in which all spins are paired; see doublet, triplet.

Single-walled carbon nanotubes (SWNT): *See* Nanotubes and buckyballs

Singularity: Defined by Vernor Vinge as the "postulated point or short period in our future when our self-guided evolutionary development accelerates enormously (powered by nanotechnology, neuroscience, AI, and perhaps uploading) so that nothing beyond that time can reliably be conceived. ...a future time when societal, scientific and economic change is so fast we cannot even imagine what will happen from our present perspective, and when humanity will become posthumanity." Another definition is the singular time when technological development will be at its fastest. A grand evolutionary leap.

Sintering: Particle coalescence of a powdered aggregate by diffusion that is accomplished by firing at an elevated temperature.

Sky Hook: A long, very strong, cable in orbit around a planet which rotates around its center of mass in such a way that when one end is closest to the ground, its relative velocity is almost zero. It would function as a kind of space elevator; shuttle craft would anchor to the end and then be lifted into orbit where they would be released. It is closely related to the idea of a beanstalk. [Originally described by Y Artsutanov in 1969. The name was propbably coined by Hans Moravec in Moravec, Hans, "A Non-Synchronous Orbital Skyhook," Journal of the Astronautical Sciences, Vol. 25, No. 4, October-December 1977, pp 307-322] [AS]

Slip casting: A forming technique used to shape ceramic materials. A slip or suspension of solid particles in water is poured into a porous mold. A solid layer forms on the inside wall as water is absorbed by the mold, leaving a shell (or a solid piece) in the shape of the mold.

Slip: Plastic deformation resulting from dislocation motion; also, the shear displacement of 2 adjacent planes of atoms.

Smart materials and products: Here, materials and products capable of relatively complex behavior due to the incorporation of nanocomputers and nanomachines. Also used for products having some ability to respond to the environment.

Smart Materials: Here, materials and products capable of relatively complex behavior due to the incorporation of nanocomputers and nanomachines. Also used for products having some ability to respond to the environment. [NTN] If you combined microscopic motors, gears, levers, bearing, plates, sensors, power and communication cables, etc., with powerful microscopic computers, you have the makings of a new class of materials: "smart materials." Programmable smart materials could shape-shift into just about any desired object.

A house made of smart materials would be quite useful and interesting. Imagine a wall changing color at your command, or making a window where their was none before. Several different types of materials exhibit sensing and actuation capabilities, including ferroelectrics (exhibiting strain in response to a electric field), shape- memory alloys (exhibiting phase transition- driven shape change in response to temperature change), and magnetostrictive materials (exhibiting strain in response to a magnetic field). These effects also work in reverse, so that these materials, separately or together, can be used to combine sensing and actuation in response to environmental conditions. They are currently in widespread use in applications from ink- jet printers to magnetic disk drives to anti- coagulant devices. An important class of smart materials is composites based upon lead zirconate titanate (PZT) and related ferroelectric materials that allow increased sensitivity, multiple frequency response, and variable frequency.

Several different types of materials exhibit sensing and actuation capabilities, including ferroelectrics (exhibiting strain in response to a electric field), shape- memory alloys (exhibiting phase transition- driven shape change in response to temperature change), and magnetostrictive materials (exhibiting strain in response to a magnetic field). These effects also work in reverse, so that these materials, separately or together, can be used to combine sensing and actuation in response to environmental conditions. They are currently in widespread use in applications from ink- jet printers to magnetic disk drives to anti- coagulant devices. An important class of smart materials is composites based upon lead zirconate titanate (PZT) and related ferroelectric materials that allow increased sensitivity, multiple frequency response, and variable frequency.

Smart polymers: Ionic gels that deform in response to electric fields. Such electro- active polymers have already been used to make "artificial muscles" (Shahinpoor et al., 1998 [147]). Currently available materials have limited mechanical power, but this is an active research area with potential applications to robots for space exploration, hazardous duty of various types, and surveillance. Hydrogels that swell and shrink in response to changes in pH or temperature are another possibility; these hydrogels could be used to deliver encapsulated drugs in response to changes in body chemistry.

Smart sensor: A sensor in which the electronics that process the output from the sensor, and forms the modifier, are partially or fully integrated on a single chip.

Smartdust: Also "Smartdust Motes" "...tiny, bottle-cap-shaped micro-machines fitted with wireless communication devices - that measure light and temperature [among other things, such as environmental monitoring, health, security, distributed processing and tracking - ed]. When clustered together, they automatically create highly flexible, low-power networks with applications ranging from climate-control systems to entertainment devices that interact with handheld computers." [What 'Smart Dust' Could Do for You By Zillah Bahar. 06/2001]

SNT: An abbreviation for structural nanotechnology; refers to integration of nanotech features into non-MNT products, also called nanomaterials.

Soft Lithography: A method for transferring a surface pattern that makes use of an elastomeric stamp.

Solvent: The component of a solution that dissolves a solute.

***sp, sp2, sp3*:** An isolated carbon atom has four valence orbitals: three mutually perpendicular *p* orbitals, each with a single nodal plane, and one spherically symmetric *s* orbital. A carbon atom in a typical molecule can be regarded as bonding with four orbitals consisting of weighted sums (termed hybrids) of these *s* and *p* orbitals. One common pattern has four equivalent orbitals, each formed by combining the three *p* orbitals with the *s* orbital; this is *sp3* hybridization. An *sp3* carbon atom forms four sigma bonds, usually in a roughly tetrahedral arrangement. Another common pattern has three equivalent orbitals formed by combining two *p* orbitals with the *s* orbital; this is termed *sp2* hybridization. An *sp2* carbon atom forms three roughly coplanar sigma bonds, usually separated by ~120 , and one pi bond (or several fractional pi bonds). If a single *p* orbital is combined with the *s* orbital, the result is *sp* hybridization, forming two sigma bonds and two pi bonds (usually in a straight line). Atoms of other kinds (e.g., N and O) can hybridize in an analogous manner.

Space Fountain: A vertical stream of magnetically accelerated pellets reaching out into space, where a station held aloft by its momentum reverses the direction and directs it towards a receiver on the ground. Essentially a simpler version of a Lofstrom loop. [I'm not sure who originated the idea, judging from Robert Forward's Indistinguishable from Magic it was a collaborative effort. A paper about the idea can be found in Hyde, Roderick A., "Earthbreak: Earth to Space Transportation," Defense Science 2003+ Vol. 4, No. 4, 1985, pp 78-92] [AS]

Span: The difference between the highest and lowest scale values of an instrument.

Species: In chemistry, a distinct kind of molecule, ion, or other structure.

Specific heat: The heat capacity per unit mass of material.

Specific Modulus (specific stiffness): The ratio of elastic modulus to specific gravity for a material.

Specific strength: The ration of tensile strength to specific gravity for a material.

Spike, The: Another term for the singularity, suggested by Damien Broderick since the growth curves look almost like a spike as it is approached. [Damien Broderick, The Spike 1997] [AS]

Spin: A quantum number describing the angular momentum and magnetic moment of a subatomic particle.

Spinning: Fiber forming process: a multitude of fibers are spun as molten material is forced through many small orifices.

Spin-Spin Coupling: Interaction between the spin states of two subatomic particles.

Spintronics: AKA: Quantum Spintronics, Magnetoelectronics, Spin Electronics. Electronic devices that exploit the spin of electrons as well as their charge. Unlike conventional electronics which is based on number of charges and their energy, and whose performance limited in speed and dissipation, spintronics is based on the direction of electron spin, and spin coupling, and is capable of much higher speed at much lower power. See our page on Spintronics. Also see Electronics and the single atom for a collection of articles (12 June 2002).

Devices that use the electron spin instead of the electron charge for electronics. Nanotech Now's Spintronics page.

Squeeze-film damping: Effect of ambient fluid and spacing on the vertical movement of a structural member with respect to a substrate.

Stability: The ability of a sensor to retain specified characteristics after being subjected to designated environmental or electrical test conditions.

Stabilizer: A polymer additive that counteracts deteriorative processes.

Stable: Strictly speaking, a system is termed stable if no rearrangement of its parts can form a system of lower free energy. In practice, the term is used with an implicit proviso regarding the transformations to be considered. Hydrogen is not considered unstable merely because it is subject to nuclear fusion at extreme temperatures. A system is usually regarded as stable (more precisely, as kinetically stable) if its rate of transformation to a state of lower free energy is negligible (by some standard) under the ambient conditions. In nanomechanical systems, a structure can commonly be regarded as stable if it has an extremely low rate of transformations when subjected to its intended operating conditions.

Standard atmosphere: A unit of pressure equal to 760 mm Hg.

Standard hydrogen electrode (SHE): A platinum conductor in contact with a 1 M H^+ ions and bathed by hydrogen gas at one atmosphere.

Star Trek scenario: Someone builds potentially dangerous self-replicating devices that spread disastrously.

State: A physical system is said to be in a particular state when its physical properties fall within some particular range; the boundaries of the range defining a state depend on the problem under consideration. In a classical world, each point in phase space could be said to correspond to a distinct state. In the real world, time-invariant systems in quantum mechanics have a set of discrete states, particular superpositions of which constitute complete descriptions of the system. In practice, broader boundaries are usually drawn. A molecule is often said to be in a particular excited electronic state, regardless of its state of mechanical vibration. In nanomechanical systems, the PES often corresponds to a set of distinct potential wells, and all points in configuration space within a particular well can be regarded as one state. Definitions of state in the thermodynamics of bulk matter are analogous, but extremely coarse by these standards.

Statistical mechanics: Statistical mechanics treats the detailed state of a system (its quantum state or, in classical models, its position in phase space) as unknown and subject to statistical uncertainties; entropy is a measure of this uncertainty. Statistical mechanics describes the distribution of states in an equilibrium system at a given temperature (describing either the distribution of probabilities of quantum states or the probability density function in phase space), and can be used to derive thermodynamic properties from properties at the molecular level. These equilibrium results are useful in nanomechanical design.

Steady-state diffusion: The diffusion condition for which there is no net accumulation or depletion of diffusing species. The diffusion flux is independent of time.

Step response: The response of a system to an instantaneous jump in the input signal.

Steric hindrance: Slowing of the rate of a chemical reaction owing to the presence of structures on the reagents that mechanically interfere with the motions associated with the reaction, typically by obstructing the reaction site.

Steric: Pertaining to the spatial relationships of atoms in a molecular structure, and in particular, to the space-filling properties of a molecule. If molecules were rigid and had hard surfaces, steric properties would merely be an opaque way of saying "shape"; a flexible side-chain, however, has definite steric properties but no fixed shape. Nanomechanical systems make extensive use of the steric properties of relatively rigid molecules, for which the term "shape" has essentially its conventional meaning so long as one remembers that the surface interactions are soft on small length-scales.

Stewart Platforms: A positional device. John Storrs info and links See also A New Family of Six Degree Of Freedom Positional Devices

Stiction: Static friction; adhering of thin micromachined layers to a substrate.

Stiffness: The stiffness of a system with respect to a deformation (e.g., the stiffness of a spring with respect to stretching) is the second derivative of the energy with respect to the corresponding displacement; this measures the curvature of the potential energy surface along a particular direction. Positive stiffness is associated with stability, and a large stiffness can result in a small positional uncertainty in the presence of thermal excitation. Negative stiffnesses correspond to unstable locations on the potential energy surface. Alternative terms for stiffness include force gradient and rigidity.

STM: A scanning tunneling microscope.

Stoichiometry: For ionic compounds, the state of having exactly the ratio of cations to anions specified by the chemical formula. Stoichiometric quantities refers to quantities of reactants mixed in exactly the correct amounts so that all are used up at the same time.

Strain (symbol å): The change in gauge length of a specimen, in the direction of an applied stress, divided by its original gauge length.

Strain gauge: An element (wire or foil) that measures a strain based on electrical resistance changes of the gauge that result from a change in length or dimension strain of the wire or foil.

Strain: In mechanical engineering, strain is a measure of the deformation resulting from stress (that is, force per unit area); the displacement of one point with respect to another, divided by their equilibrium separation in the absence of stress. In chemistry, a molecular fragment generally has some equilibrium geometry (bond lengths, interbond angles, etc.) when the rest of the molecular structure does not impose special constraints (e.g., bending bonds to form a small ring). Deviations from this equilibrium geometry are described as strain, and increase the energy of the molecule. Strain in the mechanical engineering sense causes strain in the chemical sense.

Stress concentration: The concentration or amplification of an applied stress at the tip of a notch or small crack.

Stress corrosion (cracking): A form of failure resulting from the combined action of a tensile stress and a corrosion environment, occurring at lower stress levels than required when the corrosion environment is absent.

Stress: Force per unit area applied by one part of an object to another. Pressure is an isotropic compressive stress. Suspending a mass from a fiber places it in tensile stress. Gluing a layer of rubber between two plates and then sliding one over the other (while holding their separation constant) places the rubber in shear stress.

Structural volume: The interior of a diamondoid structure typically consists of a dense network of covalent bonds; a larger excluded volume, however, is determined by nonbonded repulsions at the surface. The structural volume corresponds to a region smaller than the excluded volume, chosen to make properties such as the strength and modulus nearly size independent by correcting for surface effects.

Superconductivity: A phenomenon characterized by the disappearance of the electrical resistivity at temperatures approaching 0 K. About one third of all metals loose all electrical resistance at tempertures below a critical temperature T_c. Many elemental metals are superconductors. Some critical temperatures are: lead T_c = 7.2 K, tin T_c = 3.7 K, niobium T_c = 9.2 K, aluminum T_c = 1.2 K, mercury T_c = 4.2 K, and vanadium T_c = 5.3 K. Compounds can have even higher critical temperatures. $YBa_2Cu_3O_7$ has a critical temperature of T_c = 92 K and $HgBa_2Ca_2Cu_3O_8$ has a critical temperature of T_c = 133 K. Superconducting wires can carry currents with zero losses up to very high current densities. Above a critical current density J_c, superconductivity is supressed. A typical critical current density is 10^6 A/cm^2. Superconductivity is also supressed by strong magnetic fields.

Superconductor: An object or substance that conducts electricity with zero resistance.

Superintelligence: An intellect that is much smarter than the best human brains in practically every field, including scientific creativity, general wisdom and social skills. This definition leaves open how the superintelligence is implemented: it could be a digital computer, an ensemble of networked computers, cultured cortical tissue or what have you. It also leaves open whether the superintelligence is conscious and has subjective experiences. [Bostrom]

Superlattice Nanowire Pattern (transfer): [SNAP] a technique for producing "Ultra High Density Nanowire Lattices and Circuits". See Researchers Discover How to Make the Smallest, Most Perfect, Densest Nanowire Lattices-And It's a SNAP

Superlattice: Nanowire interwoven bundles of nanowires using substances with different compositions and properties.

Superlattices: Artificial metallic superlattices are multilayered thin films, prepared by alternately depositing two elements using vacuum deposition or sputtering techniques. A wide spectrum of elements and compounds are suitable for deposition into superlattice structures, and the range of properties displayed by the resulting superstructures is greatly dependent upon the properties of both individual lattices as well as the interaction between them. [see The Superlattice Collection for images.]

Superposition: A quantum mechanical phenomena in which a system exists in more than one state simultaneously.

Surface Plasmon: A collective motion of electrons in the surface of a metal conductor, excited by the impact of light of appropriate wavelength at a particular angle.

Surface Tension: The force acting on a liquid-gas interface resulting in thin film on the surface.

Surfactant: Short for 'SURFace ACTive AgeNT' - a surfactant is a molecule that lowers suface tension.

Synapse: A structure that transmits signals from a neuron to an adjacent neuron (or other cell).

Synthesis: The production of a specific molecular structure by a series of chemical reactions.

Synthespian: An artificial actor, for example a 3D model animated by motion capture from a real actor or a computer program. [AS]

System: In scientific usage, usually equivalent to "a collection of matter and energy being analyzed as a unit." In engineering usage, usually equivalent to "a set of components working together to serve a set of purposes."

Systematic error: An error that always occurs in the same direction.

T

T lymphocyte (T cell): Type of lymphocyte responsible for cell-mediated immunity; includes both cytotoxic T cells and helper T cells.

TAB bonding: Tape automated bonding; semiconductor packaging technique that uses a tiny lead-frame to connect circuitry on the surface of the chip to a substrate instead of wire bonds.

TATA box: Consensus sequence in the promoter region of many eucaryotic genes that binds a general transcription factor and hence specifies the position where transcription is initiated.

Technocyte: A nanoscale artificial device (especially a nanite) in the human bloodstream used for repairs, cancer protection, as an artificial immune system or for other uses. [AS 1995]

Technofobics: Those who have a phobia to technology, and/or to advances in technology.

Technology: The application of knowledge and understanding in order to control processes and fabricate products, from the Greek root *techni*, meaning 'art'.

Telomere: End of a chromosome, associated with a characteristic DNA sequence that is replicated in a special way. Counteracts the tendency of the chromosome otherwise to shorten with each round of replication. (From Greek *telos*, end.)

Telophase: Final stage of mitosis in which the two sets of separated chromosomes decondense and become enclosed by nuclear envelopes.

Temperature sensitive (ts) mutant: Organism or cell carrying a genetically altered protein (or RNA molecule) that performs normally at one temperature but is abnormal at another (usually higher) temperature.

Temperature: A system in which internal vibrational modes have equilibrated with one another can be said to have a particular temperature. Two systems A and B are said to be at different temperatures if, when brought into contact, heat flows from (say) A to B, increasing the thermal energy of B at the expense of the thermal energy of A.

Template: An external source of information or structure often used in manufacturing and mass production. As a nanoscale example, biological systems are known to use nucleic acids as a templates for protein synthesis.

Templating: The directed assembly of materials through interaction with a template.

Tensile strength (TS): The maximum engineering stress, in tension, sustainable without fracture; also called"ultimate (tensile) strength".

Terraform: To change the properties of a planet to make it more earthlike, making it possible for humans or other terrestrial organisms to live unaided on it, for example by changing atmospheric composition, pressure, temperature or the climate and introducing a self-sustaining ecosystem. This will most probably be a very long-term project, probably requiring self-replicating technology and megascale engineering. So far Venus and especially Mars looks as the most promising candidates for terraforming in the solar system. [Jack Williamson 1938] [AS] Speculation exists that with the advent of mature MNT that we should be able to accomplish Terraforming a planet such as Mars in years, rather then decades [editor]

Tertiary structure: Complex three-dimensional form of a macromolecule, especially a protein.

Tesla [T]: Unit of magnetic induction: 1T = 1 weber/m^2 (also, 1T = 10^4 gauss).

Tetrahymena: Genus of ciliated protozoa used in studies of ciliary axonemes, self-splicing RNA, and telomere reproduction.

TFT: Thin film transistor.

The Atom: Atoms are the building blocks of matter. All things are made of arrangements of atoms. There are 118 (with some very short lived ones) different elements in the periodic table, which means that we know of 118 different kinds of atoms.

Thermal conductivity (ê): For steady-state heat flow, the proportionality constant between the heat flux and the temperature gradient. Also, a parameter characterizing the ability of a material to conducting heat.

Thermal energy: The internal energy present in a system

as a result of the energy of thermally equilibrated vibrational modes and other motions (including both kinetic energy and molecular potential energy). The mean thermal energy of a classical harmonic oscillator is kT.

Thermal expansion coefficient: The rate of change of length with respect to temperature for a particular material.

Thermal expansion coefficient, linear (á): The fractional change in length divided by the change in temperature.

Thermal fatigue: A type of fatigue failure that introduces the cyclic stresses by fluctuating thermal stresses.

Thermal fluctuations: The thermal energy of a system (or of a particular part or mode of motion in a system) has a mean value determined by the temperature and by the structure of the system. Statistical deviations about that mean are termed thermal fluctuations; these are of great importance in determining both rates of chemical reactions and error rates in nanomechanical systems.

Thermal Noise: the vibration and motion of atoms and molecules caused by the fact that they have a temperature above absolute zero. [RCM] Once used as an argument on why MNT could not work. Since refuted: See Thatís impossible! How good scientists reach bad conclusions

Thermal shock: The fracture of a brittle material resulting from stresses introduced by a rapid temperature change.

Thermal stress: A residual stress introduced within a body resulting from a change in temperature.

Thermistor: A temperature-measuring device, that contains a resistor or semiconductor whose resistance varies with temperature.

Thermocouple: A temperature-measuring device, which contains a pair of end-joined dissimilar conductors in which an electromotive force is developed by thermoelectric effects when the joined ends and the free ends of the conductors are a different temperature.

Thermocycler: An instrument that repeatedly cycles through various temperatures required for an iterative, temperature-dependant chemical process such as the polymerase chain reaction.

Thermodynamics: A field of study embracing energy conversion among various forms, including heat, work, and potential and kinetic energy.

Thermoelastic: Both stress and temperature changes alter the dimensions of an object having a finite stiffness and a nonzero thermal expansion coefficient. Applying a stress then produces a temperature change; this can result in a heat flow which then changes the stress: these are thermoelastic effects, and result in losses of free energy.

Thermogenic limit: In medical nanorobotics, the maximum amount of waste heat that may safely be released by a population of in vivo medical nanorobots that are operating within a given tissue volume.

Thermoplastic polymer: A substance that when molded to a certain shape under appropriate conditions can later be remelted. A substance that when molded to a certain shape under pressure and high temperatures cannot be softened again or dissolved.

Thioester bond: High-energy bond formed by a condensation reaction between an acid (acyl) group and a thiol group (- SH); seen, for example, in acetyl CoA and in many enzyme-substrate complexes.

Thiol: An SH group, or a molecule containing one. Also known as a sulfhydryl or mercapto group.

Thiol: *See* sulfhydryl

Threshold: The smallest input signal that will cause a readable change in the output signal.

Thrombin : The enzyme that converts *fibrinogen* to *fibrin* to form a blood clot. Thrombin circulates in the inactive form, prothrombin.

Thylakoid: Flattened sac of membrane in a chloroplast that contains pigment and carries out the light-gathering reactions of photosynthesis. Stacks of thylakoids form the grana of chloroplasts.

Tight junction: Cell-cell junction that seals adjacent epithelial cells together, preventing the passage of most dissolved molecules from one side of the epithelial sheet to the other.

Tight-receptor structures: A receptor structure in which a bound ligand of a particular kind is confined on all sides by repulsive interactions (note that favorable binding energies are compatible with repulsive forces). A tight-receptor structure discriminates strongly against all molecules larger than the target.

Time constant: The time it takes for the output change to reach 63% of its final value.

Tissue Engineered Medical Products Standards TEMPS: The standards process for tissue engineered medical products is underway within the American Society for Testing and Materials (ASTM)

Tissue engineering: Generating tissue *in vitro* for clinical applications, such as replacing wounded tissues or impaired organs. A *cell culture* facility is required for cell harvest and expansion.

The term "tissue engineering" was coined at an NSF [National Science Foundation] -sponsored meeting in 1987. At a later NSF- sponsored workshop, tissue engineering was defined as "...the application of principles and methods of engineering and life sciences toward fundamental understanding ...and development of biological substitutes to restore, maintain and improve [human] tissue functions." This definition is intended to include procedures where the biological substitutes are cells or combinations of different cells that may be implanted on a scaffold such as natural collagen or as synthetic, biocompatible polymers to form a tissue.

Tissue models: Cells, tissues, and organs function in a 3-D environment. Ideally, cell-based models that capture both the 3-D organization and multicellular complexity of the target system provide the most powerful tools for screening the effects of therapeutic candidates. This meeting will bring together tissue engineers who are constructing the 3-D models with biologists who are studying healthy vs. diseased states and pharmacologists who are developing high throughput screening assays. As with 3-D biological models, each specialty provides insight into the complete system.

Top Down Molding: [AKA: mechanical nanotechnology] Carving and fabricating small materials and components by using larger objects such as our hands, tools and lasers, respectively. [NTN] Opposite of Bottom Up.

Top-down fabrication: Building a small structure carving it out of a macroscopic piece of material.

Top-down nanotechnology: Engineers taking existing devices, such as transistors, and making them smaller. Top- down or mechanical nanotechnology will have the greatest impact on our everyday lives in the near future.

Top-Down: Molding, carving and fabricating small materials and components by using larger objects such as our hands, tools and lasers.

Topoisomerase (DNA topoisomerase): Enzyme that makes reversible cuts in a double helical DNA molecule for the purpose of removing knots or unwinding excessive twists.

Toughness: A measure of the amount of energy absorbed by a material as it fractures, indicated by the total area under the material's tensile stress-strain curve.

Transcription: Conversion of genetic information from DNA to messenger RNA (mRNA).

Transduction (self-generating or modulating): The conversion of the signal to be measured into another, more easily accessible form. Source of energy for transmission of the sensor signal.

Transduction mode (direct or indirect): How the sensor acquires the desired information from the material. In general, this parameter is an indication of the ability of the sensor signal to provide information regarding a material property or state of interest.

Transformer: Device using magnetically linked inductors to change AC voltage level.

Transhuman: Someone actively preparing for becoming

posthuman. Someone who is informed enough to see radical future possibilities and plans ahead for them, and who takes every current option for self-enhancement.

Transhumanism: Philosophies of life (such as Extropianism) that seek the continuation and acceleration of the evolution of intelligent life beyond its currently human form and human limitations by means of science and technology, guided by life-promoting values. [Max More 1990]

Transient response: The response of the sensor to a step change in the measurand.

Transistor: Semiconductor device used for amplification and switching. The basic element in an integrated circuit. An on/off switch (consisting of three layers of a semiconductor material) that consists of a source (where electrons come from), a drain (where they go) and a gate that controls the flow of electrons through a channel that connects the source and the drain. There are two kinds of transistor, the bipolar transistor (also called the junction transistor), and the field effect transistor.

Transition state theory: Any of several theories that give approximate descriptions of chemical reaction rates based on the PES of the system, and in particular, on the properties of two potential wells and a transition state between them.

Transition state: At the saddle point of a col linking two potential wells, the direction of maximum negative curvature defines the reaction coordinate; the transition state is a hypothetical system of reduced dimensionality, free to move only on a hypersurface perpendicular to the reaction coordinate at its point of maximum energy.

Translation: Conversion of genetic information from messenger RNA (mRNA) into protein.

Transmission electron microscope (TEM): A microscope that produces an image by using electron beams to transmit (pass through) the specimen, making examination of internal features at high magnifications possible. A method of producing images of a sample by illuminating the sample with electronic radiation (in a vacuum), and detecting the electrons that are transmitted through the sample.

Transmission: Refers to system for carrying electric power at voltages above 100,000 volts.

Transtegumental: Crossing or passing through the skin or covering of a body.

Tribology: Study of friction, wear and lubrication of interacting surfaces. [BNL]

Tribology: The science and technology of two interacting surfaces in relative motion and of related subjects and practices. The popular equivalent is friction, wear, and lubrication in surfaces sliding against each other, as in bearing and gears.

Triple bond: A double bond is formed when a pi bond is superimposed on a single bond; adding a second pi bond results in a triple bond. The two pi bonds have perpendicular nodal planes, and their sum has roughly cylindrical symmetry, permitting rotation in much the same manner as a single bond.

Triplet: An electronic state of a molecule in which two spins are aligned. This term is derived from spectroscopy: a system of two aligned spins has three possible orientations with respect to a magnetic field; each has a different energy, resulting in sets of three field-dependent spectral lines (see doublet, singlet.)

TST: Transition state theory.

Tubeologist: Someone who knows their nanotubes inside and out, such as David Tom·nek [uhf]

Tunneling: A classical particle or system could not penetrate regions in which its energy would be negative, that is, barrier regions in which the potential energy is greater than the system energy. In the real world, however, a wave function of significant amplitude may extend into and beyond such a region. If the wave function extends into another region of positive energy, the barrier is crossed with some probability; this process is termed tunneling (since the barrier is penetrated rather than climbed).

Turing Test: Turing's proposed test for whether a machine is conscious (or intelligent, or aware): we communicate via text with it and with a hidden human. If we can't tell which of our partners in dialogue is the human, we say the computer is conscious. [AS]

Turn-on-voltage: Applied voltage required to produce conduction in a diode.

U

Ubiquitin: Small, highly conserved protein present in all eucaryotic cells that becomes covalently attached to lysines of other proteins. Attachment of a chain of ubiquitins tags a protein for intracellular proteolytic destruction in a proteasome.

Ubiquitous Computing: Also known as "embodied virtuality", "smart environment" and "ambient intelligence". Computers that are an integral, invisible part of people's lives. In some ways the opposite of virtual reality, in which the user is absorbed into the computational world. With ubiquitous computing, computers take into account the human world rather than requiring humans to enter into the computer's methods of working. [AS] See our Smartdust page.

ULSI: Ultra large scale integration; a chip with over 1,000,000 components.

Unimolecular: Occurring to or within a single molecule; like intramolecular, but can refer to fragmentation reactions.

Unit cell: The basic structural unit of a crystal structure, defined in terms of atom (or ion) positions within a parallelepiped volume.

Universal Assembler: This is a term that was introduced by Eric Drexler. In his terminology, an assembler is enzyme or catalyst that brings two molecules together and connects them. DNA can be used to program the assembly of a wide variety of proteins. Drexler imagines a universal assembler to be something that could be programmed to fabricate anything. The following quote was taken from Eric Drexler's book Engines of creation in the section where he discusses universal assemblers.

These second-generation nanomachines - built of more than just proteins - will do all that proteins can do, and more. In particular, some will serve as improved devices for assembling molecular structures. Able to tolerate acid or vacuum, freezing or baking, depending on design, enzyme-like second-generation machines will be able to use as "tools" almost any of the reactive molecules used by chemists - but they will wield them with the precision of programmed machines. They will be able to bond atoms together in virtually any stable pattern, adding a few at a time to the surface of a workpiece until a complex structure is complete. Think of such nanomachines as assemblers.

Because assemblers will let us place atoms in almost any reasonable arrangement (as discussed in the Notes), they will let us build almost anything that the laws of nature allow to exist. In particular, they will let us build almost anything we can design - including more assemblers. The consequences of this will be profound, because our crude tools have let us explore only a small part of the range of possibilities that natural law permits. Assemblers will open a world of new technologies.

Advances in the technologies of medicine, space, computation, and production - and warfare - all depend on our ability to arrange atoms. With assemblers, we will be able to remake our world or destroy it. So at this point it seems wise to step back and look at the prospect as clearly as we can, so we can be sure that assemblers and nanotechnology are not a mere futurological mirage.

Uses raw atoms and molecules to construct consumer goods, and is pollution free. Can be programmed to build anything that is composed of atoms and consistent with the rules of chemical stability. Eric Drexler talks about these assemblers as nanorobots with telescoping manipulator arms that are capable of picking up individual atoms, and combining them however they are programmed.

Universal Constructor: A machine capable of constructing anything that can be constructed. The physical analog of a "universal computer", which can perform any computation. [AS]

Unsaturated: Describes a molecule that contains one or more double or triple carbon-carbon bonds, such as isoprene or benzene. Possessing double or triple bonds.

Uplift: To increase the intelligence and help develop a culture of a previously non- or near-intelligent species. [From the Uplift novels by David Brin] [AS]

Upload: (a) To transfer the consciousness and mental structure of a person from a biological matrix to an electronic or informational matrix (this assumes that the strong AI postulate holds). The term "Downloading" is also sometimes used, mainly to denote transferring the mind to a slower or less spacious matrix. (b) The resulting infomorph person. [The origin of the term is uncertain, but obviously based on the computer technology term 'uploading' (loading data into a mainframe computer).] [AS]

URL: Universal Resource Locator — address of a World Wide Web site.

Usenet: Interlinked bulletin boards available via Internet and commercial on-line services.

Utility Fog: [AKA: Polymorphic Smart Materials] Objects formed of "intelligent" polymorphic (able to change shape) substances, typically having an octet truss structure. Concept concieved by Dr. J. Storrs Hall.

UV: Ultraviolet — characterization of short-wavelength light for exposing photoresist in making semiconductor devices.

V

Vacuum: An environment with pressure lower than that of the standard atmosphere (14.7 PSI or 760 Torr).

Valence band: The electron energy band that contains the valence electrons in solid materials.

Valence electrons: Electrons that can participate in bonds and in chemical reactions; lone-pair electrons are valence electrons, although not participating in a bond. The electrons in the outermost occupied electron shell, that participate in interatomic bonding.

Valence: In covalent compounds, the valence of an atom is the number of bonds it forms to other atoms.

Van der Waals bond: A secondary, permanent or induced, interatomic bond between adjacent molecular dipoles. Any of several intermolecular attractive forces not resulting from ionic charges; in this volume, only London dispersion forces are described by this name. Descriptions of the potential energy of nonbonded interactions follow the convention of expressing van der Waals attractive forces and overlap repulsion forces as a single *van der Waals potential*.

Van der Waals forces: Any of the noncovalent attractive or repulsive forces acting upon neutral atoms or molecules.

Vasculocyte: In medical nanorobotics, a theorized (nanorobotic) device capable of performing repairs of an injured vascular luminal surface.

Vasculography: A physical description (and mapping) of the human vascular system.

Vasculoid: Is a single, complex, multisegmented nanotechnological medical robotic system capable of duplicating all essential thermal and biochemical transport functions of the blood, including circulation of respiratory gases, glucose, hormones, cytokines, waste products, and cellular components.

The vasculoid [concept] is a single, complex, multisegmented nanotechnological medical robotic system capable of duplicating all essential thermal and biochemical transport functions of the blood, including circulation of respiratory gases, glucose, hormones, cytokines, waste products, and cellular components.

Vasculoid is a single, complex, multisegmented nanotechnological medical robotic system capable of duplicating all essential thermal and biochemical transport functions of the blood, including circulation of respiratory gases, glucose, hormones, cytokines, waste products, and cellular components. See *Vasculoid: A Personal Nanomedical Appliance to Replace Human Blood.*

VC (Venture capitalist): A person or organisation that invests in promising companies during early development stages.

Vector: In cell biology, an agent (virus or plasmid) used to transmit genetic material to a cell or organism. (See also cloning vector.)

Vegetal pole: The end at which most of the yolk is located in an animal egg. The end opposite the animal pole.

Ventral: Situated toward the belly surface of an animal.

Vesicle: Small, membrane-bounded, spherical organelle in the cytoplasm of a eucaryotic cell.

Virtual Nanomedicine: Using VR to perform surgery and other functions inside the body.

Virtual reality system: A combination of computer and interface devices (goggles, gloves, etc.) that presents a user with the illusion of being in a three dimensional world of computer-generated objects.

Virus: A parasite (consisting primarily of genetic material) that invades cells and takes over their molecular machinery in order to copy itself. A small replicator consisting of little but a package of DNA or RNA which, when injected into a host cell, can direct the cell's molecular machinery to make more viruses. Particle consisting of nucleic acid (RNA or DNA) enclosed in a protein coat and capable of replicating within a host cell and spreading from cell to cell. Often the cause of disease.

Viscoelasticity: A type of deformation exhibiting the mechanical characteristics of viscous flow and elastic deformation.

Viscosity: The ratio of the magnitude of an applied shear stress to the velocity gradient that it produces; in other words: a measure of a noncrystalline material's resistance to permanent deformation.

Vitamins (engineering): In machine replication theory, vitamin parts are components of a self-replicating machine which the machine is incapable of producing itself, therefore these vital parts must be supplied from an external source.

Vitrification: During firing of a ceramic body, the formation of a liquid phase that becomes a glass-bonding matrix upon cooling.

VLSI: Very large scale integration; a chip with 100,000 to 1,000,000 components.

V_{max}: Maximum rate of an enzymatic reaction when a substrate is present at saturation levels.

Volitional normative model of disease: In medical nanorobotics, disease is said to be present in a human being upon either (1) the failure of optimal physical (e.g. biological) functioning, or (2) the failure of desired (by the patient) functioning.

Volt: Unit of electrical potential difference (see"voltage").

Voltage [V]: Potential difference between two points: energy to move a 1-C charge through a 1-V potential difference is 1-J.

Vombinatorial biology: Involves genetic manipulation of bacteria and fungi that produce complex natural products. This technology includes construction of large libraries of recombinant microbes capable of generating novel organic molecules and engineering secondary metabolite biosynthetic pathways to modify valuable biologically active microbial metabolites. *Combinatorial materials design:* Uses computing power (sometimes together with massive parallel experimentation) to screen many different materials possibilities to optimize properties for specific applications (e.g., catalysts, drugs, optical materials).

Von Neumann Machine: (pronounced von noi-man) A machine which is able to build a working copy of itself using materials in its environment. This is often proposed as a cheap way to mine or colonize the entire solar system or galaxy. An early fictional treatment was the short story "Autofac" by Philip K. Dick, published in 1955, which actually seems to precede John von Neumann's original paper about self-reproducing machines (von Neumann, J., 1966, The Theory of Self-reproducing Automata, A. Burks, ed., Univ. of Illinois Press, Urbana, IL.). [AS]

VON Neumann Probe: A von Neumann Machine able to move over interstellar or interplanetary distances and to utilize local materials to build new copies of itself. Such probes could be used to set up new colonies, perform megascale engineering or explore the universe. [AS]

Vone substitutes: Synthetic or natural materials for the replacement of bones or bone tissue. They include hard tissue replacement polymers, natural coral, hydroxyapatite, beta- tricalcium phosphate, and various other biomaterials. The bone substitutes as inert materials can be incorporated into surrounding tissue or gradually replaced by original tissue.

possible applications (e.g. retail) [illegible] organized materials.

W

Wafer: Semiconductor disk out of which integrated circuits are made (also see"chip","mask").

Watt (W): Unit of power. One watt is the power that, in 1 second, gives rise to an energy of 1 joule.

Wave function: In quantum mechanics, a complex function extending over the configuration space of a system; its complex conjugate yields the probability density function, and other mathematical operations yield other physical quantities.

Weber: Unit of magnetic flux. One weber is a magnetic flux that, linking a circuit of 1 turn, would produce in it an electromotive force of 1 volt if it were reduced to zero at a uniform rate in 1 second.

Weight percent (wt%): Concentration specification on the basis of weight (or mass) of a particular element relative to the total alloy weight (or mass).

Western blotting: Technique by which proteins are separated and immobilized on a paper sheet and then analyzed, usually by means of a labeled antibody.

Wet Nanotechnology: The study of biological systems that exist primarily in a water environment. The functional nanometer-scale structures of interest here are genetic material, membranes, enzymes and other cellular components. The success of this nanotechnology is amply demonstrated by the existence of living organisms whose form, function, and evolution are governed by the interactions of nanometer-scale structures. [Rice University]

Whisker: A very thin, single crystal of high perfection which has an extremely large length-to-diameter ratio. Whiskers are used as the reinforcing phase in some composites.

White biotechnology: An emerging field within modern biotechnology that serves industry. It uses living cells like moulds, yeasts or bacteria, as well as enzymes to produce goods and services. Living cells can be used as they are or improved to work as "cell factories" to produce enzymes for industry.

White blood cell (leucocyte): Nucleated blood cell lacking hemoglobin; includes lymphocytes, neutrophils, eosinophils, basophils, and monocytes.

Wild-type: Normal, nonmutant form of an organism; the form found in nature (in the wild).

Work: Energy transferred by applying a force over a distance; lifting a mass does work against gravity, and stores gravitational potential energy.

World Wide Web (WWW): Graphical hypertext system linking many Internet computers.

X

Xenografts: The live cells, tissues and organs used in xenotransplant procedures.

Xenopus laevis (South African clawed toad): Species of frog (not a toad) frequently used in studies of early vertebrate development.

Xenotransplantation: Xenotransplantation involves the use of live animal cells, tissues and organs in the treatment or mitigation of human disease. Xenotransplantation has the potential to provide an alternative source of tissue for human transplantation. However, there are concerns that in crossing the species barriers, infectious agents which normally do not cause any problem in the original species may be changed or may adversely respond to the new species into which the infectious agent is transplanted.

X-ray crystallography: Technique for determining the three-dimensional arrangement of atoms in a molecule based on the diffraction pattern of x-rays passing through a crystal of the molecule.

Y

Yeast: Common term for several families of unicellular fungi. Includes species used for brewing beer and making bread, as well as pathogenic species (that is, species that cause disease).

Yield strength: The stress required to produce a very slight yet specified amount of plastic strain; a strain offset of 0.002 is commonly used.

Yielding: The onset of plastic deformation.

Young's modulus: A modulus relating tensile (or compressive) stress to strain in a rod that is free to contract or expand transversely in accord with its Poisson's ratio. The relevant measure of strain is the elongation divided by the initial length.

Z

Z disc (Z line): Platelike region of a muscle sarcomere to which the plus ends of actin filaments are attached. Seen as a dark transverse line in micrographs.

Zener diode: Semiconductor diode that has a well-defined turn-on voltage for conduction in the reverse direction.

Zeptosecond: One-billion-trillionth of a second, or 10^{-21} second. Because nuclear movement takes place so quickly, scientists would need a pulse of light lasting just one zeptosecond to observe them. Johns Hopkins University

Zero offset: The output of a sensor at zero input for a specified supply voltage or current.

Zero Sum: A term used to describe a situation in which one entity can gain only if other entities suffer an equal loss; for example, a private poker game.

Zettatechnology: In which zetta means 10^{21}, referring to the typical number of distinct designed parts in a product made by the systems we envision (molecular, mature, or molecular-manufacturing-based nanotechnology). The term refers to the implemented technology and its products, rather than to intermediate steps on the pathway.

Zinc finger: Structural motif seen in many DNA-binding proteins, composed of a loop of polypeptide chain held in a hair-pin bend bound to a zinc atom.

Zippocytes: In medical nanorobotics, a theorized medical nanorobot that can rapidly perform incision-wound repairs to the dermis and epidermis; dermal zippers.

Zone refining: A metallurgical process for obtaining a highly pure metal that depends on continuously melting the impure material and recrystallizing the pure metal.

Zygote: Diploid cell produced by fusion of a male and female gamete. A fertilized egg.